AF539717

# Diseases of Field & Horticultural Crops and Their Management-II

NIPA GENX ELECTRONIC RESOURCES & SOLUTIONS P. LTD.
New Delhi-110 034

STARTUP INDIA # DIPP14488

# NIPA GENX ELECTRONIC RESOURCES & SOLUTIONS P. LTD.

Publishers | Library Suppliers | Subscription Agents | Exporters and Importers

**www.nipaers.com**

**An Inbuilt Dictionary Supporting 13 Languages Like Hindi, Malayalam, Tamil, Telugu, Kannada Marathi, Konkani, Odia, Bangla, Gujarati, Punjabi, Urdu, Sanskrit.**

***Browse***, **Search**
***Read & Buy***

- ✓ Print Books
- ✓ Ebooks
- ✓ E Chapters
- ✓ New Titles
- ✓ Forthcoming
- ✓ Subject Catalogs

***Online Tutorial Resources On***

- ✓ Current Affairs
- ✓ Reasoning, Logic & Aptitude
- ✓ Competitive Examinations
- ✓ English Language Learning
- ✓ Personality Development
- ✓ Mock Interviews

**Pay Using**

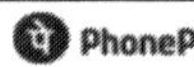

paytm PayPal UPI CCAvenue

STARTUP INDIA # DIPP14488

**NIPA GENX ELECTRONIC RESOURCES & SOLUTIONS P. LTD.**
101,103 Vikas Surya Plaza, CU Block, LSC Market, Pitam Pura, New Delhi 110034
Ph: +91 11 - 27341616, 27341717, 27341718
Email: newindiapublishingagency@gmail.com
Website: www.nipaers.com

## About The Author

**Dr. Sanjeev Kumar** is Assistant Professor/Scientist, Department of Plant Pathology, Jawaharlal Nehru Krishi Vishwavidyalaya, Jabalpur, Madhya Pradesh. He is very sincere and hard working teacher. He has taught many undergraduate, postgraduate and Ph.D courses with great dedication to the great satisfaction of the students and actively involved in all teaching activities. Dr. Kumar has guided fifteen M.Sc and one Ph.D student and has been actively involved in his research and extension activities too. As a researcher, he has been associated with 8 major research projects sponsored by ICAR, NATP,TSP and JICA, Japan. He has published 68 research and reviews papers, 25 book chapters, 105 popular articles, 1 research bulletin and 3 technical folders. He is an active member of over six National and International Societies. He is the fellow member of Indian Phytopathological Society,IARI New Delhi, Indian Society of Mycology and Plant Pathology,Udaipur, Rajasthan and Society of Biocontrol Advancement, Banguluru. He has attended many conferences, symposia, workshops and presented papers. He has also attended the World Soybean Research Conference held in Durban, South Africa.

Dr. Kumar has authored six books entitled Plant Pathogens & Principles of Plant Pathology, Diseases of Horticultural Crops: Identification & Management, Diseases of Field Crops and Their Integrated Management, Pesticides and Plant Protection Appliances, Fundamentals of Plant Pathology published by New India Publishing Agency, Pitam Pura, New Delhi and Textbook of Diseases of Field & Horticultural Crops & Their Management-I, published by Brillion Publishing, Desh Bandhu Gupta Road, Karol Bagh, New Delhi – 110005 and three practical manuals. His research interests include Pulse pathology and biological control

# Diseases of Field & Horticultural Crops and Their Management-II

**Sanjeev Kumar**

Assistant Professor/Scientist
Department of Plant Pathology
Office of Dean, Faculty of Agriculture
Jawaharlal Nehru Krishi Vishwa Vidyalaya
Jabalpur-482004, Madhya Pradesh, India

**NIPA GENX ELECTRONIC RESOURCES & SOLUTIONS P. LTD.**
New Delhi-110 034

**NIPA GENX ELECTRONIC RESOURCES & SOLUTIONS P. LTD.**

101,103, Vikas Surya Plaza, CU Block
L.S.C.Market, Pitam Pura, New Delhi-110 034
Ph : +91 11 27341616, 27341717, 27341718
E-mail:newindiapublishingagency@gmail.com
www: www.nipabooks.com

***For customer assistance, please contact***
Phone: + 91-11-27 34 17 17 Fax: + 91-11- 27 34 16 16
E-Mail: feedbacks@nipabooks.com

© 2022, Publisher

ISBN: 978-93-91383-91-6

All rights reserved. No part of this publication may be reproduced, stored in a retrieval system or transmitted in any form or by any means, including electronic, mechanical, photocopying recording or otherwise without the prior written permission of the publisher or the copyright holder.

This book contains information obtained from authentic and highly reliable sources. Reasonable efforts have been made to publish reliable data and information, but the author/s, editor/s and publisher cannot assume responsibility for the validity, accuracy or completeness of all materials or information published herein or the consequences of their use. The work is published with the understanding that the publisher and author/s are not attempting to render any professional services. The author/s, editor/s and publisher have attempted to trace and acknowledge the copyright holders of all material reproduced in this publication and apologize to copyright holders if permission and/or acknowledgements to publish in this form have not been taken. If any copyrighted material has not been acknowledged, please write to us and let us know so that we may rectify the error, in subsequent reprints.

**Trademark Notice:** NIPA, the NIPA logos and their presentations (the way they are written/ presented) in this book are the trademarks of the publisher and hence may not be used without written permission, if copied or used without authorization, the infringer will be prosecuted as per law.

NIPA also publishes books in a variety of electronic formats. Some content that appears in print may not be available in electronic books, and vice versa.

Composed and Designed by NIPA.

**Dedicated**
**to**
**My Parent**

# Preface

This book of Diseases of Field Crops & their Management-II is a carefully written for undergraduate class and is completely in accordance with the latest 5th Deans's Committee Recommendations Guidelines.

The book aims to develop in the young minds the passion for learning and analysing the host, pathogen and environment, to probe into the problems and think of the possible solution. While preparing the book, I have made a conscious effort to address the needs of students at various levels and have graded our content accordingly.

This book has some features of its own, These are:

**Fun Time**: Sets the tone by addressing general terminology which also relate to the subject. Fun time is good ice breaker and prepares the students for the information to follow.

**Infohive:** Stimulates the young minds as they look to learn, yet newer things in, compare and broaden their horizons.

**Text:** The text in the book is clear, well presented, sequencially arranged. Each chapter has further be divided into sub-heads such as important diseases and their causal organisms, diagnostic symptoms on different parts, etiology, disease cycle, favorable conditions and integrated managment. The chapters are designed in a way that leads to the comprehensive learning of the key concepts, help in the development of the investigative skill of the students.

I do not claim originality in the preparation of this book and has taken help from a large number of books, journals, periodicals, bulletins, google etc. I humbly thank authors, editors, and publishers of the books, journals, etc. I am also indebted to my wife Dr. (Mrs) Archana Rani for inspiration and help and children Saumya and Adyan for bearing with me during the preparation of this book.

We expect the book to set the students on a pursuit where they learn more by asking thant by being told. It would be our pleasure to incorporate suggestions and improvements, proposed by teachers and students, in our future editions.

**Sanjeev Kumar**

August, 2021
Jabalpur

# Contents

# List of Colour Plates

# Key Terms (Fun Time)

| Sl.No | Term | Definition |
|---|---|---|
| **Symptom and Sign** | | |
| 1 | Symptom | Abnormal plant growth or function due to a disease. A symptom is a reaction of the plant. |
| 2 | Sign | Pathogen parts or products seen on a host plant. A sign is the pathogen itself. |
| 7 | Blight | Sudden drying and browning of whole leaves, shoots or branches. |
| 8 | Canker | A typically well-developed canker is a symptom complex exhibiting necrotic, atrophic and hypertrophic symptoms. |
| 9 | Chlorosis | Yellowing of green tissue due to chlorophyll destruction or failure of chlorophyll formation. |
| 10 | Curl | Abnormal bending or curling of leaves or shoots due to localized overgrowth on one side or in certain tissues. |
| 11 | Damping-off | A symptom complex characterized by rapid dying, browning, and rotting of germinating seedlings. Shoots may be killed before emergence, stems may be attacked in the root collar region causing shoots to fall over, roots may be destroyed, or cotyledons may be attacked. |
| 12 | Defoliation | Loss of leaves through abscission. |
| 13 | Dieback | Progressive drying, shrivelling, and browning of twigs or branches from the tips inward toward the trunk. |
| 14 | Dwarfing | Subnormal size of a plant or some of its organs. |
| 15 | Gall | A pronounced tumefaction, often more or less spherical and usually composed of undifferentiated cells. |
| 16 | Girdling | Tangential enlargement of a canker or lateral coalescence of cankers causing a branch or stem to be encircled and resulting in the complete stoppage of conduction. |
| 17 | Gummosis | Formation of gums by diseased cells and tissues and the extrusion of gum from wounds and other lesions. |
| 22 | Knot | A type of tumefaction. |
| 23 | Leaf retention | Abnormally long retention of leaves usually resulting from a failure to develop the abscission meristem. |
| 24 | Lesion | A localized area of diseased tissue. |
| 25 | Mosaic | Pale green mottling of leaves. |

| Sl.No | Term | Definition |
|---|---|---|
| 29 | Oozes | Viscid masses composed of living or dead pathogen structures and partially disintegrated host tissues. |
| 32 | Scab | A limited, more or less circular, raised, and sometimes roughened lesion on fruits, tubers, leaves, and stems resulting from an overgrowth of epidermal, cortical, and peridermal tissues. |
| 39 | Tumor | Local swelling on any part of the plant, usually woody roots, stem, or branches, usually resulting from stimulation of the plant meristem by the pathogen. |
| 40 | Wart | A horny, hardened protuberance. |
| 41 | Wilting | A flaccid appearance of leaves and shoots resulting from a temporary or permanent loss of turgor due to excess transpiration by the leaves and shoots. |
| 42 | Yellowing | Loss of green color from chlorophyllous tissues, due to the destruction of the chlorophyll . |
| 43 | Ooze | Viscid masses composed of living or dead pathogen structures and partially disintegrated host tissues. |
| 44 | Cyst | A female nematode body filled with eggs produced by some nematodes. |
| 48 | Sclerotia | Macroscopic mass of hyphae, usually rounded and darkened. |
| 50 | Spore | Reproductive propagule of fungi and oomycetes. |
| 51 | Pustule | Small blister like elevation of epidermis created as spores from underneath and push outward. |
| **Oomycetes** | | |
| 52 | Hypha | Vegetative structure of fungi and oomycetes. |
| 53 | Coenocytic hyphae | Hyphae without crosswalls, found in oomycetes |
| 54 | Oospore | Sexual spore produced by oomycetes characterized by a thick cell wall, this is the resting or survival stage. |
| 55 | Oogonium | Female gamete of oomycete that is fertilized by an antheridium to form an oospore. |
| 56 | Antheridium | Male gamete of oomycete that fertilize an oogonium to from an oospore. |
| 57 | Sporangium | Non-motile asexual spore produced by oomycetes, lemon or globulose in shape. |
| 58 | Sporangiophore | A specialized hyphae that produces sporangia. |
| 59 | Zoospore | A motile asexual spore characterized by two flagella. |
| 60 | Germ tube | The early growth of mycelium from a germinating spore. |
| **Ascomycetes and Deuteromycetes** | | |
| 61 | Septate | Having cross walls in a hypha or spore. A cross wall is called a septum. |
| 62 | Germ tube | Early growth of of a hypha produced by a germinating fungus spore. |
| 63 | Apressorium | Swollen tip of a hypha or germ tube that facilitates attachment and penetration of the host by a fungus. |

| Sl.No | Term | Definition |
|---|---|---|
| 64 | Haustorium | A specialized fungal hyphae that enters and absorbs nutrients from a host cell. |
| 65 | Ascomycetes | Fungi that produce sexual spores in sac-like structures called asci. |
| 67 | Anamorph | Imperfect or asexual stage of a fungus. |
| 68 | Teleomorph | Perfect or sexual stage of a fungus. |
| 69 | Conidia | Asexual, non-motile spores of fungi. |
| 70 | Conidiophore | A specialized hypha that produces conidia. |
| 74 | Apothecia | An open cup-shaped ascocarp. |
| 75 | Perithecia | A flask-shaped ascocarp with an opening for releasing spores. |
| 76 | Cleistothecia | An ascocarp where the asci are completely enclosed. |
| 77 | Sporodochium | An asexual fruiting structure consisting of a cluster of conidiophores woven together on a mass of hyphae. |
| 78 | Pycnidium | A flask-shaped asexual fruiting body with an opening for releasing spores.. |
| 79 | Acervulus | A subepidermal, saucer-shaped, asexual fruiting body producing conidia on short conidiophores. |
| 80 | Synnema | An asexual fruiting body consisting of fused conidiopores to form a stalk with conidia on the end. |
| 81 | | A thick-walled asexual spore formed by the modification of a hyphal cell. |
| 82 | Sclerotia | Macroscopic mass of hyphae, usually rounded and darkened. |
| **Basidiomycetes** | | |
| 86 | Basidiospores | Sexual spores of basidiomycetes prodcued on basidia. |
| 90 | Teliospores | Overwintering spores of rusts and smuts that produce basidia and basidiospores. |
| 91 | Aecia | Rust fruiting bodies that produce aeciospores. |
| 92 | Aeciospores | Dikaryotic spore of a rust fungus produced in an aecium; in heteroecious rusts, a spore stage that infects the alternate host. |
| 96 | Uredia | Rust fruiting bodies that prodcue uredospores. |
| 97 | Uredospores | Asexual, dikaryotic, often rusty-colored spore of a rust fungus, produced in a structure called a uredinium; the "repeating stage" of a heteroecious rust fungus, i.e. capable of infecting the host plant on which it is produced. |
| 98 | Macrocyclic | Rusts that produce all five spore types. |
| 99 | Microcyclic | Rusts that lack one or more of the five spore types. |
| 100 | Heteroecious | Rusts that require two different hosts to complete their life cycle. |
| 101 | Autoecious | Rusts that complete their life cycle on a single host. |
| 103 | Resting spore | Long term survival spores produced by plasmodiophoromycetes. |
| **Bacteria** | | |
| 106 | Plasmid | Small, circullar, extrachromosomal DNA found in bacteria. |
| 107 | Gram positive | Group of bacteria with one membrane in association with their cell wall. |

| Sl.No | Term | Definition |
|---|---|---|
| 108 | Gram negative | Group of bacteria with two membranes in association with their cell wall. |
| 109 | Fission | Bacterial reproduction by simple cell division. |
| **Viruses** | | |
| 110 | Virus | A submicrscopic obligate parasite consisting of nucleic acid and protein. |
| 111 | Viroids | Small RNAs that can infect a host and cause disease |
| 115 | Coat Protein | Protein, encoded by virus nucleic acid, which covers the viral nucleic acid |
| 116 | Inclusion body | Aggregation of virus particles in a host cell that is visible with a compound microscope |
| 117 | Plasmodesmata | A connection across a plant cell wall that connects the cytoplasm of two neighboring cells |
| 118 | Aquisition period | Period of time need for a virus vector to aquire a virus while feeding on an infected plant host. |
| 119 | Inoculation period | Period of time need for a virus vector to transmit a virus while feeding on an uninfected plant host. |
| 120 | Latent period | Period of time before a vector is infective after picking up the virus from an infected host. |
| 122 | Mechanical transmission | Virus transmission from plant to plant by infected plant sap |
| 123 | Vector transmission | Transmission of a pathogen from plant to plant by another organism |
| **Nematode** | | |
| 124 | Stylet | A long, slender, hollow feeding strcuture used by plant parasitic nematodes. |
| 125 | Ectoparasite | A parasite that feeds from outside the host. |
| 126 | Endoparasite | A parasite that enter and feed from inside the host. |
| 127 | Sedentary | Life style of plant parasitic nematodes that stay in one place and set up a feeding site. |
| 128 | Migratory | Life style of plant parasitic nematodes that move through the host as they feed. |
| **Epidemiology** | | |
| 129 | Epidemic | A disease increase in a population; usually a widespread and severe outbreak of disease. |
| 130 | Epidemiology | Study of factors affecting the outbreak and spread of disease. |
| 131 | Disease incidence | Number or poroportion of individuals that are diseased. |
| 132 | Disease severity | Amount or proportion of tissue that is diseased |
| 133 | Monocycic disease | A disease where only one disease cycle is completed each year or growing season. |
| 134 | Polycyclic disease | A disease with several cycles each year of inoculum production and infection. |

| Sl.No | Term | Definition |
|---|---|---|
| 135 | Primary inoculum | Inoculum that produces the first infection of plants in a year or growing season. |
| 136 | Secondary inoculum | Inoculum produced by previous infections that infect in the same year or growing season. |
| 137 | Inoculum density | Number of infective units in a given volume or area. |
| **Resistant** | | |
| 147 | Susceptible | Phenotypic expression related to extensive symptom development and/or pathogen reproduction and accomplished by uninhibited invasion of host by pathogen. |
| 148 | Resistant | Phenotypic expression related to complete or partial suppression of symptom severity and/or pathogen reproduction and accomplished by arrested or slowed invasion of host by pathogen. |
| 150 | Pathogenicity/ Virulence | Ability of a microbe to cause disease (invade, infect, cause symptoms, reproduce). |
| 151 | Race | A genetically and often geographically distinct mating group within a species; also a group of pathogens that infect a given set of plant varieties. |
| 152 | Aggressiveness | Virulent forms of pathogen cause differing degrees of symptom severity. |
| **Chemical** | | |
| 153 | Preventative | Chemical control method aimed at preventing infection of the pathogen. |
| 154 | Curative | Chemical control method aimed at inhibiting the development of an established infection. |
| 155 | Protectant | Chemical meant to reside on the plant surface as a preventative control measure. |
| 156 | Systemic | Chemical meant to be taken up by and distributed throughout the plant as a preventative or curative control measure. |
| 157 | ED 50 | Effective dose, the amount needed to have the desired effect in 50% of the population. |
| 158 | Active ingredient | In pesticides, the chemical responsible for the desired effect. |
| 159 | Mode of action | Molecluar mechanism of a pesticide; how the chemical interacts with the pathogen. |
| 160 | Fumigant | A toxic gas or volitile substance that is used to disinfest soil of various pests. |

# Infohive (Lets Compare)

## 1. Disease and Disorder

| Characters | Disease | Disorder |
|---|---|---|
| Definition | Any deviation in the general health, or physiology or function of plant or plant parts, is recognized as a disease | Plant disease in which no pathogens or parasites is associated with the cause is known as Non parasitic or non infectious or physiological disorders. |
| Symptom development | Complex symptoms, possibly both primary and secondary symptoms may not point out directly to causal agent (e.g. wilt of plants; could be a result of root rot, vascular plugging, a non- infectious disorder such as water deficiency or by a combination) | Relatively simple; possibly limited to one, or only primary symptoms often pointing rather directly causal agents (e.g. sun burn certain plants derived of sufficient shade at one location). |
| Symptom expression | Appear progressively at definite stages; individual symptoms may reach full intensity gradually or quickly | Appear suddenly, nearly once in their fullest intensity |
| Signs | Present | Absent |
| Plant species affected | Wide range of plant species affected | Only certain species are affected |
| Distribution of affected plants | Usually irregular or uneven distribution | Fairly regular or uniform in field or tightly clustered in affected area with no apparent spread pattern |

## 2. Symptom and Sign

| Characters | Symptom | Sign |
|---|---|---|
| Definition | External and internal reaction or alterations of a plant as a result of disease. | Structure of the pathogen itself. |
| Diagnostic | Not always true for diagnostic to study disorder | Help in identifying the pathogen and diagnosis of the disease |
| Examples | Leaf spot, Blight, Wilt, Rot, Yellowing, Blossom end rot | Spores, Sclerotia, Apothecia, Hyphae, Fruiting body |

## 3. Parasite and Pathogen

| Characters | Parasite | Pathogen |
|---|---|---|
| Definition | Phenomenon of the growth of one organism, the parasite, at the expense of another, the host. | An entity, usually a micro organism that can cause the disease. |
| Plant- pathogen Relationship | All parasites are not pathogens | Most (not all) pathogens are parasites |
| Examples | Root nodule bacteria *Rhizobium leguminoserum* on the roots of pulses.Mycorrhizal fungus parasitic on roots of trees | *Phytophthora infestans* causes late blight in potato |

## 4. Infection and Infestation

| Characters | Infection | Infestation |
|---|---|---|
| Definition | Initiation and establishment of a parasite within a host plant | Presence in numbers (e.g., of insects, mites, or nematodes). Do not confuse with "infection," a term that applies only to living, diseased plants or animals. |
| Relation with plant | May cause diseases only when environmental condition is favorable | Directly related to plant diseases |

## 5. Pythium and Phytophthora

| Features | Pythium | *Phytophthora* |
|---|---|---|
| Papilla | Absent | Present |
| Haustorium | Absent | Present |
| Sporangium | Globose to oval sporangia are present either terminal or intercalary on the somatic hyphae | Developed on specialized hyphae |
| Vesicle | Present | Absent |
| Zoospore differentiation | Yes | No |
| Hyphal wall | Contains little amount of protein | Contains higher amount of protein |

## 6. Peronospora and Sclerospora

| Characters | Sclerospora | Peronospora |
|---|---|---|
| Mycelium | Profusely branched and intercellular | Well developed and intercellular |
| Haustorium | Small and vesicular | Short and knob like or filamentous and more or less branched |
| Conidiophore | Are erect, solitary or in groups of 2 or 3 | Consists of an erect trunk which is branched 2-10 times and branches are more or less on acute angles |

| | | |
|---|---|---|
| Conidia | Conidia are elliptical, hyaline and smooth | Hyaline to slightly tinted but they do not have papillae |
| Germination | Conidia germinate by forming germ tubes when they are true conidia | Spores- are true conidia germinate by germ tubes |
| Germination through zoospore formation | Yes when they are true sporangia | Never |
| Oospore | Globose intra mycelial with a brown and irregularly wrinkled epispore | Globose or ellipsoidal, reticulate , tuberculate |

## 7. Ustilago and Sphacelotheca

| **Characters** | **Ustilago** | **Sphacelotheca** |
|---|---|---|
| Sori | Naked without false membrane | Composed of central columella |
| Spore coloration | Olivaceous brown in color that looks lighter on one side and darker on the other | Dark brown in mass when seen |
| Shape of spore | Spherical to Oval | Round to slightly |
| Spiny | Epispore has fine spines | No spines |

## 8. Fusarium and Colletotrichum

| **Characters** | **Fusarium** | **Colletotrichum** |
|---|---|---|
| Fruiting body | Sporodochia | Acervuli |
| Conidia | Macroconidia are long, falcate,fusiform, septate, Microcnidia single celled , crescent shaped, | Conidia are hyaline and ovate not in large groups |

## 9. Helminthosporium, Alternaria and Cercospora

| **Characters** | **Alternaria** | **Helminthosporium** | **Cercospora** |
|---|---|---|---|
| Shape of conidia | Muriform and obclavate | Fusoid to elongate or clavate | Vermiform and obclavate niddle shaped |
| Tips of conidia | Conidia are beaked | Both the ends of conidia are round | Both the end pointed |
| Number of conidia | Conidium are in chains in acropetal succession | Usually single conidium is found on conidiophores at one time | Multicelled conidia |
| Conidiophore | Non fasciculate and usually arise singly | Fasciculate | Variable , well developed , branched and simple |
| Shape of conidiophores | Straight | Knee joint type | Straight |

## 10. Viroid and Virusoid

| Characters | Viroid | Virusoid |
|---|---|---|
| Type of RNA | Small, low molecular weight RNA that themselves and cause diseases | Low molecular weight , circular linear RNA molecules and absence of bifurcates |
| Sequence homology | With virioids | Absence sequence homology with virusoid |
| Protein | Present | Absent |
| Examples | Potato spindle tubers | Solanum nodiliform mottle virus (RNA) |

## 11. Gram Positive Bacteria and Gram Negative Bacteria

| Characters | Gram Positive | Gram Negative |
|---|---|---|
| Reaction to Gram's | Appear as violet or blue or a red background | Appear themselves stained red |
| Teichoic acid | Present | Absent |
| Sensitivity to Penicillin | Highly Sensitive | Resistant |
| Acid fast staining | Occassional | Never |
| Magnesium protein | Present | Absent |
| Examples | Bacillus, Streptomyces | Xanthomonas, Pseudomonas |

## 12. Monocyclic and Polycyclic diseases

| Characters | Monocyclic Disease | Polycyclic Disease |
|---|---|---|
| Definition | A disease where only one disease cycle is completed each year or growing season. | Some pathogens specially a fungus, can complete a number of life cycles within one crop season of the host plant and the disease caused by such pathogens is called multiple cycle disease |
| Rate of Increase | Like simple interest in money | Like compound interest in money |
| Reproduction Cycle | Long (Reproduce only once during crop | Short (Reproduce many times during crop) |
| Birth rate | Low | High |
| Death rate | Low | High |
| Inoculums Survival | Long lived | Short lived |
| Examples | Soil borne diseases | Air borne diseases |

## 13. Necrotroph and Biotroph

| Characters | Necrotroph | Biotroph |
|---|---|---|
| Definition | A pathogenic fungus that kills the host and survives on the dying and dead cells. | A plant pathogenic fungus that requires living host cells i.e. an obligate parasite. |
| Rate of killing host | Slow | Relatively slow |

| | | |
|---|---|---|
| Haustoria formation | Absent | Present |
| Mode of host penetration | Absent | Haustoria present |
| Host | Wide | Narrow |
| Stages of | Infect juvenile, Adult at senescence stages | All stage of host development |
| Host- pathogen contact | Can grow saprophytically from the host | Necessary |
| Examples | Sclerotium, Rhizoctonia | Viruses and pathogens causing, Rusts, Mildews, Powdery mildews |

## 14. Powdery Mildew and Downey Mildew

| **Characters** | **Powdery Mildew** | **Downey Mildew** |
|---|---|---|
| Definition | Powdery growth often on plant surface especially on leaves | Downy growth with angular spots on a plant surface. |
| Type of symptom | Necrosis | Chlorosis |
| Leaf part affected | Mostly appear on both upper and lower leaves | Mostly appear on lower leaves |
| Type of mycelium | True | False |
| Mycelium | Septate | Aseptate |
| Mobility of Asexual Spore | Non motile | Motile |
| Sexual spore | Ascospore, Clistothecium | Oospore |
| Fungal growth composition | Sporangiophore and sporangia | Conidiophore and conidia |
| Examples | *Erysiphe, Uncinula, Leveillula, Podosphaera, Phyllactinia* | *Peronospora, Plasmopara* |

## 15. Rust and Smut

| **Characters** | **Rust** | **Smut** |
|---|---|---|
| Perfect spore | Terminal | Intercalary |
| Sporidial number | Four | Indefinite |
| Mechanisms of discharge | Active | Passive |
| Basidiocarp | Absent | Rare |
| Parasitism | Obligate | Facultative |
| Clamp connections | Rare | Common |
| Alternation of generation | Distinct | Indistinct |

## 16. Necrosis and Apoptosis

| **Characters** | **Necrosis** | **Apoptosis** |
|---|---|---|
| Cell volume | Increases | Decreases |
| Integrity of plasma membrane | Damaged plasma membrane rupture | Plasma membrane remains intact up to the moment of formation of apoptotic vesicles |

| Cell | Appear into intercellular spaces | Split into spaces |
|---|---|---|
| State of nucleus | Nucleous destroyed | Nuclear fragment |
| Organelles | Swollen and destroyrd | Pore open into outer to liberate apoptosis inducers |

## 17. Plasmogamy and Karyogamy

| Characters | Plasmogamy | Karyogamy |
|---|---|---|
| Definition | Process of union of two protoplasts | Process of fusion of two nuclei which are brought together by plasmogamy |
| Location of | Two nuclei are brought together closer with in the same cell | Both the nuclei takes place within the same cell |
| Zygote or Oospore formation | No such formation | Zygote of oospore having 2N number of chromosome is formed |
| Phase of sexual reproducation | First phase | Second phase |

## 18. Amphigynous and Paragynous

| Characters | Amphigynous | Paragynous |
|---|---|---|
| Definition | Oogonium grows through the antheridium | Antheridium itself within the oogonium |
| Location of | Oogoium and antheridium develop on different hphal tips | Oogonium and antheridium develop from the branch of the same hyphae or from and adjacent branch |
| Examples | *Phytophthora infestans* | *Phytophthora cactarum* |

## 19. Sexual cycle and Parasexual cycle

| Characters | Sexual Cycle | Parasexual Cycle |
|---|---|---|
| Nuclear fusion | Takes place in structure resulting in hybrid zygotes | Nuclear fusion is very rare in vegetable cells. Heterozygotes may be detected by color and nutrition. Homozygotes rarely formed and not detected |
| Zygote persistence | Only one nuclear generation | Many mitotic divisions |
| Recombination and crossing | Recombination at meiosis; crossing over at tetrad stage in all the chromosome pair | Recombination by rare accidents of two waves (I) mitotic crossing over at 4 tetrad stage and haplodization probably *via* annuploidy |
| Recognition of recombinant products | Rarely recognized isolated | Recombination occurs among vegetative cells. Usually recognized by suitable markers |

## 20. Hypertrophy and Hyperplasia

| Characters | Hypertrophy | Hyperplasia |
|---|---|---|
| Definition | A symptom due to an abnormal increase in the size of individual cells. | A symptom due to an abnormal increase in the number of individual cells. |
| Examples | Galls, Canker, Witches broom | Scab |

## 21. Collateral and Alternate host

| Characters | Alternate host | Collateral host |
|---|---|---|
| Definition | One of two kinds of plants on which a parasitic fungus (e.g., rust) must develop to complete its life cycle. | Wild host of same families of a pathogen is called as collateral host. |
| Host family | Belongs to different family | Belongs to same family |
| Examples | Wheat, Barbery is an alternate host of *Puccinia graminis tritici* causing black stem rust. | Rice, *Drechslera oryzae* can survive through their weed hosts like *Echinocloa colorum and Setaria intermedia* |

## 22. Localized and Systemic infection

| Characters | Localized Infection | Systemic Infection |
|---|---|---|
| Definition | Developed by the infection of a pathogen | Occurs due to spreading throughout the plant body internally |
| Damages | Cause lesser damage than systemic infection of the host | Cause severe damage to the host plant and entire plant may be killed |
| Pathogen movement | Movement restricted to the infected part | Pathogen grom from the point of entry of vearying extents without showing adverse effects on tissue thorugh which it passes |
| Examples | Leaf spot, Blight | Bacterial wilt |

## 23. Disease and Life Cycle

| Characters | Disease Cycle | Life Cycle |
|---|---|---|
| Definition | Chain of events involved in disease development. | Stages or successive stages in the growth and development of an organism that occur between appearance and reappearance the same stage (i.e. spore) organism |

## 24. Virulence and Aggressiveness

| Characters | Virulence | Aggressiveness |
|---|---|---|
| Definitions | Degree of pathogenicity | Ability of a race to grow and invade to cause disease on large scale in susceptible plants |
| Component of Pathogenicity | Qualitative | Quantitative |
| Use of term | Often in relation to plant | No, used in relation to pathogen |
| Parasitism | Facultative | Obligate |
| Clamp connections | Common | Rare |
| Alternation of generation | Indistinct | Distinct |

## 25. Haustoria and Appresoria

| Characters | Haustoria | Appresoria |
|---|---|---|
| Defnition | Special branch of fungus especially intercellularly within living cell to absorbs nutrients | Swollen tip of hyphae that facilitates attachment and penetration of host cells |
| Function | Sucking nutrition from host attachment | Penetration of host cell |
| Examples | VAM fungi, Downy and Powdery mildew | Most of the fungi |

## 26. Paraphysis and Periphysis

| Characters | Paraphysis | Periphysis |
|---|---|---|
| Definition | Sterile hypha present in some ostiole of fruiting bodies of fungi intermingled a perithecium as in pyrenomycetes | Sterile hypha present in some spore bearing structure and lie in the opening of an ascostroma as in along the asci in a hymenium .Loculoascomycetes |

## 27. Autoecious and Heteroecious rust

| Characters | Autoecious Rust | Heteroecious Rust |
|---|---|---|
| Definition | Rust pathogen which complete its life cycle one host | Requiring two or more unrelated hosts for completing the life cycle of a rust. |
| Alternate host | Not Required | Required |
| Examples | Linseed rust (*Melampsora lini*), Bean rust(*Uromyces* spp.) | Stem rust (Wheat &Barley) Coffee rust |

## 28. Oospore and Oosphere

| Characters | Oospore | Oosphere |
|---|---|---|
| Definition | Thick walled resting spore formed from a fertilized oogonium | Large naked spherical female gamete (egg) |
| Motility | Motile | Non-motile |
| Sexuality | Sexual spore | Female gamete |

## 29. Asci and Basidium

| Characters | Asci | Basidium |
|---|---|---|
| Phylum | Ascomycotina | Basidiomycotina |
| Description | Sexual fruiting body which contain usually 8 ascospores | Club shaped structure that usually contains 4 basidiospores |
| Site | Ascospores are produced on ascus | Basidiospores are produced on basidium |
| Septation | Aseptate | Both septate and aseptate |
| Fruiting body | Ascocarp | Basidiocarp |

## 30. Sporodochium and Acervulus

| Characters | Acervulus | Sporodochium |
|---|---|---|
| Definition | A saucer-shaped, spore-producing body of a fungus embedded in host tissue. | A cushion-shaped spore-producing body of a fungus. |
| Composition | Conidiophores are short and arise from close crowded hyphae. Conidia are intermingled | Consists of two layer upper and lower. The lower part is a cushion storm like mass of hyphae which breaks through the host tissue from the exposed upper surface conidia bearing conidiophores arises. It constitutes the upper part of sporodochia |
| Family | Melanconiaceae | Tuberculariaceae |
| Genus | *Colletotrichum*<br>*Pestalotia* | *Fusarium*<br>*Graphium* |

## 31. Pycnidium and Perithecium

| Characters | Pycnidium | Perithecium |
|---|---|---|
| Description | Flask shaped asexual fruiting body | Small rounded or .flask shaped fruiting body |

## 32. Apoplastic and Symplastic movement

| Characters | Apoplastic | Symplastic |
|---|---|---|
| Synonymous term | Acropetal movement | Basipetal movement |
| Direction of movement | Upward | Downward |
| Movement after absorption | Systemic fungicides after their absorption in plants move in the direction of evapo transpirtion | Systemic fungicides after their absorption in plants move photosynthate towards the sink |
| Energy | Active | Passive |
| Examples | Benomyl, Carbendazim | Fosetyl Al |

## 33. Protectant and Systemic fungicide

| Characters | Protect ant Fungicide | Systemic (Penetrant) Fungicide |
|---|---|---|
| Site of action | Multisite | Single site |
| Nature of Protection | Prophylactic | Curative |
| Movement after absorption | Many metabolic system of fungal pathogens are affected | Few metabolic system of fungal pathogens are affected |
| Phytotoxic effect | Common | Rare |
| Resistant Development | Rare | Common |
| Action on plant system | Localized | Whole plant system |
| Distribution on plant system | Confined in the plant where it applied | Trans located inside the plant |

# DISEASES OF FIELD CROPS

# 1

# Diseases of Wheat and Their Management

| Sl.No | Major Diseases | Causal Organism |
|---|---|---|
| 1 | Loose smut | *Ustilago tritici* (Pers.) E. Rostr. |
| 2 | Black or stem rust | *Puccinia graminis tritici* Pers |
| 3 | Brown or leaf rust | *Puccinia recondita* f.sp. *tritici* Eriks. & Henn |
| 4 | Yellow or stripe rust | *Puccinia striiformis* var. *striiformis* Westend |
| 5 | Flag smut | *Urocystis agropyri (G. Preuss)* J. Schröt. |
| 6 | Hill bunt or Stinking smut | *Tilletia tritici (syn. Tilletia caries)* Bjerk. Wint. and *T. laevis (syn. T. foetida).* Kuhn |
| 7 | Karnal bunt | *Tilletia indica* Mitra |
| 8 | Powdery mildew | *Blumeria graminis (DC.)* Speer |
| 9 | Leaf blight | *Alternaria triticina* Prasada & Prabhu |

## Rust

The rust of wheat has been studied more than any other disease in the world. In India, K.C Mehta and his associates studied this disease with exemplary zeal. Three rusts of wheat are known. These are Black stem rust, caused by *Puccinia graminis tritici*, Brown rust (leaf rust), caused by *Puccinia recondita*, and Yellow rust, caused by *P. striiformis*. Brown and yellow rusts cause more damage thant the black rust, because of early appearance in the season.

## Black or Stem Rust

*Puccinia graminis tritici*

### Diagnostic Symptoms

- Symptoms are produced on almost all aerial parts of the wheat plant but are most common on stem, leaf sheaths.
- Brown pustules of uredina appear on lower surface of the leaves, leaf sheaths and the stems, and on the spikes. which give them a rusty appearance.
- Uredina form urediniospores , which are the repeating spores.
- Uredina gradually start producing teliospores and finally transform into telia.

- Telia are black and crust like, formed abundantly on the stem.
- For this reason, the rust is also called black stem rust.
- Other stymptoms are stunted growth of the plants and poor tilleirng , which in severe infections grains, are not formed (Fig. 2).

**Etiology**

**Systemic position**

| | |
|---|---|
| Phylum | Basidiomycotina |
| Class | Teliomyetes |
| Order | Uridinales |
| Family | Pucciniaceae |
| Genus | *Puccinia* |
| Species | *gramins tritici* |

- Pathogen is an biotroph and has a complex life cycle featuring alternation of generations.
- Pathogen is heteroecious, requiring two hosts to complete its life cycle - the wheat and barberry .
- Wheat is the primary host and barberry is the alternate host.
- There are many species in *Berberis* and *Mahonia* that are susceptible to stem rust, but the common barberry is considered to be the most important alternate host.
- *P. graminis* can complete its life cycle either with or without barberry (the alternate host).
- *P. graminis* is macrocyclic (exhibits all five of the spore types that are known for rust fungi).
- Characteristic rust color on stems and leaves is typical of a general stem rust as well as any variation of this type of fungus.

**Disease Cycle**

**On Wheat**

- Due to its cyclical nature, there is no true 'start point' for this process. Here, the production of urediniospores is arbitrarily chosen as a start point.
- Urediniospores are formed in structures called uredinia, which are produced by fungal mycelia on the cereal host 1–2 weeks after infection.

- Urediniospores are dikaryotic (contain two un-fused, haploid nuclei in one cell) and are formed on individual stalks within the uredinium. They are spiny and brick-red.
- Urcdiniospores are the only type of spores in the rust fungus life cycle which are capable of infecting the host on which they are produced and this is therefore referred to as the 'repeating stage' of the life cycle.
- It is the spread of urediniospores which allows infection to spread from one cereal plant to another. This phase can rapidly spread the infection over a wide area.
- Towards the end of the cereal host's growing season, the mycelia produce structures called telia.
- Telia produce a type of spore called teliospores.These black, thick-walled spores are dikaryotic.
- Teliospores are the only form in which *Puccinia graminis* is able to overwinter independently of a host.
- Each teliospore undergoes karyogamy (fusion of nuclei) and meiosis to form four haploid spores called basidiospores. This is an important source of genetic recombination in the life cycle.
- Basidiospores are thin-walled and colourless.

**Barberry**

- Basidiospores cannot infect the cereal host, but can infect the alternative host (Usually barberry).They are usually carried to the alternative host by wind.
- Once basidiospores arrive on a leaf of the alternative host, they germinate to produce a haploid mycelium which directly penetrates the epidermis and colonises the leaf.
- Once inside the leaf the mycelium produces specialised infection structures called pycnia.
- Pycnia produce two types of haploid gametes, the pycniospores and the receptive hyphae.
- Pycniospores are produced in a sticky honeydew which attracts insects. The insects carry pycniospores from one leaf to another.
- Splashing raindrops can also spread pycniospores.
- A pycniospore can fertilize a receptive hypha of the opposite mating type, leading to the production of a dikaryotic mycelium.
- This is the sexual stage of the life cycle and cross-fertilisation provides an important source of genetic recombination.

- This dikaryotic mycelium then forms structures called aecia, which produced a type of dikaryotic spores called aeciospores.
- These have a worty appearance and are formed in chains - unlike the urediniospores which are spiny and are produced on individual stalks.
- Chains of aeciospores are surrounded by a bell-like enclosure of fungal cells.
- Aeciospores are able to germinate on the cereal host but not on the alternative host (they are produced on the alternative host, which is usually barberry).
- They are carried by wind to the cereal host where they germinate and the germ tubes penetrate into the plant.
- Fungus grows inside the plant as a dikaryotic_mycelium. Within 1–2 weeks the mycelium produces uredinia and the cycle is complete.

**Disease cycle without barberry**

- Since the urediniospores are produced on the cereal host and can infect the cereal host.
- It is possible for the infection to pass from one year's crop to the next without infecting the alternative host (barberry).For instance, infected volunteer wheat plants can serve as a bridge from one growing season to another.
- In other cases the fungus passes between winter wheat and spring wheat, meaning that it has a cereal host all year round.
- Since the urediniospores are wind dispersed, this can occur over large distances.
- This cycle consists simply of vegetative propagation - urediniospores infect one wheat plant, leading to the production of more urediniospores which then infect other wheat plants

**Favorable Conditions**

- Optimum temperatures above - $20^0$C.
- Availability of free water.
- Deposition of dew on leaf surface.

**Integrated Management**

- Use of resistant varieties like HP2278, HW741, WL614, Sonara 63,64 etc. Mostly ,the resistant cultivar used in India contain the Sr31 gene, which, unfortunately is susceptible to the fast spreading UG-99 race of the pathogen. The race originally came up in Uganda in 1999, and is fast spreading.It may soon enter South Asia, and knock at our door. The wheat breeders have been altered by GRI (Global Rust Initiative) regarding the need for developing cultivars resistant to the UG-99.

- Mixed cropping of wheat and barley with suitable crop.
- Reduction in proportion of nitrogen in the N.P.K in a fertilizer.
- Seed treatment with Carboxin or Oxycarboxin @ 2.5 g/kg seed.
- Foliar spray with Propiconazole @ 0.1 % or Pyraclostrobin 133g/L+Epoxiconazole 50g/L @ 0.1%. if required repeat after 15days.

## Annual recurrence of rust in India

- *Puccinia graminis tritici* is a heterocious, macrocyclic rust.
- Hoewever, in India and many other countries the alternate host, i.e. barberry , donot play any role in the recurrence of rust.
- Late Prof. K.C.Mehta has been the pioneer worker in connection with annual recurrence of rust in India.
- According to him. Only urediniospores are accountable for the initiation and spread of rust from year to year.
- *Barberis* does not play any active role as an alternate host.
- The aecial stages present on barberry in India actually belong to *Aecidium montanum* and not to *Puccinia graminis tritici.*
- Teliospores do not survive at temperature over $26^0$C.
- In the Indian plains, the teliospores on wheat are produced in February-March. After march the day temperature in plains is always more than $26^0$C.
- The urediniospores are the chief source of the rust infection on wheat in India. But urediniospores also cannot survive the high summer temperature of Indian plains.
- Mehta investigated and proved that these are those urediniospores that over summer in the northern hills at higher altitudes of 1300-2500 m where they survive on self sown wheat plants and tillers.
- The conditions of high altitudes and low temperature on hills are favorable for the survival of urediniospores. Hence they retain their vitality in hills.
- These surviving urediniospores of hills first infect the wheat crop near or at the foot hills, where they are carried easily by the air currents.
- From these infected wheat plants in the foot hills they are carried by the winter winds to the plains, where they infect the wheat crop in January-February.

- Mehta therefore suggested that the rust severity in the plains can be reduced if there is no wheat cultivation on the hills for some time.

## Brown or Leaf Rust

*Puccinia triticina (P. recondita)*

- This is the most cmmon rust in North and South India.
- It appears earlier than the black rust.

## Diagnostic Symptoms

- Most common site for symptoms is on upper leaf blades, however, sheaths, glumes and awns may occasionally become infected and exhibit symptoms.
- Symptoms begin as small, circular to oval yellow spots on infected tissue of the upper leaf surface.
- As the disease progresses, the spots develop into orange colored pustules which may be surrounded by a yellow halo.
- The pustules produce a large number of spores that are easily dislodged from the pustule resulting in an "orange dust" on the leaf surface or on clothes, hands and equipment.
- As the disease progresses, black spores may be produced resulting in a mixture of orange and black lesions on the same leaf.
- Tiny orange lesions may be present on seed heads, but these lesions do not develop into erumpent pustules (Fig. 3).
- Yield loss often occurs.

## Etiology

## Systemic position

Kingdom : Fungi

Phylum :Basidiomyotina

Class :Teliomycetes

Order :Uredianles

Family :Pucciniaceae

Genus :*Puccinia*

Species :*recondita*

- Pathogen is biotroph and heteroecious.
- Primary host is wheat and the secondary host is *Thalictrum.*
- Five types of spores are formed in the life cycle.
- Uredospores, teleutospores, and basidiospores develop on wheat plants and pycnidiospores and aeciospores develop on the alternate hosts.
- Uredospores are brown, round or oblong, echinulate, stalked and have 7-10 germ pores.These spores are dispersed in the air and cause infection to other plants.
- Teliospores are bicelled with flattened top of the upper cell, brown and smooth.They are produced by teliosori, which are small, oval to linear, black in color, and covered by epidermis.
- Teliospores germinate to produce basidia and basidiospores, which normally cause infection in alternate host.
- Functional alternate host, which has not been reported from India, is *Isopyrum fumarioides.*
- In India, the non functional host is *Thalictrum,* which commonly occurs in hills.
- Spermogonial and aecial stages of the pathogen have not been reported.

**Disease Cycle**

- Pathogen over-summers at altitudes of 1450-2050m on Nilgiri and Pulney hills in South and Himalayas in North India.
- Wind blown urediniospores dispersed from site of survival and rust get established in early january in plain of Tamil Nadu and Karnataka in south, and in the foot hils of the Himalayas in the north.
- Uredostores of north and south moves in opposite directions, Finally merging into each other. This may be result in epidemics in any part of conutry if climatic condition becomes favourable.

  The first build up of inoculum takes place in the plain of Karnataka in south and uredospores and wind blown north wards to Maharashtra and Madhya Pradesh.
- They germinate and penetrate through closed stomata on the lower surface, and form dikaryotic mycelium.
- Uredinia are formed and urediniospores cause secondary infections, till end of the season when teliospores are formed.
- The pycnial and aecial stages are not reported in India.

## Favorable Conditions

- Relative humidity- 100% .
- Optimum temperature - 20–25 °C.
- Free moisture

## Integrated Management

- Collect and destroy crop debris.
- Provide irrigation at critical stages of the crop.
- Avoid water logging.
- Avoid water stress during flowering stage.
- Balance application of NPK.
- Follow mixed cropping and crop rotation
- Avoid excess application of "nitrogen".
- Foliae spray with Propiconazole 25% EC @ 0.1 percent, if required repeat after fifteen days.

## Yellow or Stripe Rust

***Puccinia striiformis***

- Yellow rust does considerable damage in Punjab, Haryana and Uttar Pradesh, which are the main wheat growing areas.
- It is rare in South India because of the high heat during the crop season.

## Diagnostic Symptoms

- Mainly occur on leaves than the leaf sheaths and stem.
- Bright yellow pustules (Uredia) appear on leaves at early stage of crop and pustules are arranged in linear rows as stripes.
- Stripes are yellow to orange yellow and this gives the name 'yellow' rust to the disease.
- In severe infections, all parts of the plant are infected, and the stripes disappear due to the crowding of uredinia.
- Teliospores are also arranged in long stripes, dull black in colour.
- Telia appear at the end of the season, as black streaks, on the lower surface.

## Etiology

### Systemic position

| | |
|---|---|
| Phylum | Basidiomycotina |
| Class | Teliomyetes |
| Order | Uredinales |
| Family | Pucciniaceae |
| Genus | *Puccinia* |
| Species | *Striiformis* |

- Pathogen is biotroph and its primary host is wheat.
- Uredospores of rust pathogen are almost round or oval in shape and bright orange in colour.
- Each uredospores possesses 6-10 scattered germ pores.
- Teliospores are bright organge to dark brown, two celled and flattened at the top.
- Sterile paraphyses are also present at the end of sorus.
- No intervening hosts for pycnial and aecial stages of the pathogen have been discovered.

### Disease Cycle

- Inoculum survives in the form of uredospores /teliospores in the hills during off season on self sown crop or volunteer hosts, which provide an excellent source of inoculum.
- About ninety weeds serve as collateral hosts, which can provide the urediniospores for primary infection.
- Wind borne uredospores from hills are lifted due to cyclonic winds cause primary infection in the plains during crop season.
- In India, role of alternate host (Barberis) is not there in completing the life cycle.
- Urediniospores spread the disease until the end of the season, when telia are formed.
- Teliospores have no role in disease causation.

### Favorable Conditions

- Low temperature -10-15°C
- Relative humidity- >60 percent

### Integrated Management

- Grow resistant varieties like PBW 343, PBW 550, PBW 17.
- Collect and destroy crop debris.
- Provide irrigation at critical stages of the crop.
- Avoid water logging.
- Avoid water stress during flowering stage.
- Balance application of NPK.
- Follow mixed cropping and crop rotation.
- Avoid excess application of "N".
- Foliar spray with Propiconazole 25% EC @ 0.1 percent, if required repeat after fifteen days.

**A comparison of the three wheat rusts.**

| | Leaf rust | Stem rust | Stripe rust |
|---|---|---|---|
| Pathogen | *Puccinia triticina (P. recondita)* | *Puccinia graminis tritici* | *Puccinia striiformis* |
| Month of appearance | January | March-April | December-January |
| Pustule location | Leaf, mainly on the upper surface | Stem and leaf, upper and lower surfaces of leaf; occasionally on head and seeds | Leaf, upper surface; occasionally on head and seeds |
| Pustule color | Orange-brown | Orange-red to dark-red | Orange-yellow |
| Pustule arrangement | Single and random | Single and random | Stripes |
| Pustule shape and size | Round or slightly elongated; small to medium | Oval shaped or elongated; small to large | Round, blister-like; small |
| Tearing of host epidermis | Rare | Conspicuous | None |
| Uredospores | Spherical, brown, echinulate, wall possesses 7-10 germ pores, 16-20 mm in diameter | Oval, echinulate, brick red or rusty in appearance, posses thick cellulose wall containing four germ pores situated in an equatorial , 25-30x17-20µm | Spherical to ovate, wall colorless and finally echinulate possessing 6-16 germ pores, vary variable in size from 23-35 by 20-35µm |
| Teliospores | Resemble those of yellow rust in shape and size, top cell flattened at apex, sorus divided into small chambers by paraphyses | Thick walled, smooth, bicelled, super-imposed cells, the top cell being rounded or thickened at the apex, dark brown and measure 40-60x15-20µm | Dark brown, bicelled, the top cell flattened at the apex, in contact with the epidermis, interspersed with brown and unicellular paraphyses measure 35-63x12-24µm |
| Optimum temperature for infection | 59-68°F | 59-84°F | 45-54°F |
| Optimum temperature for disease development | 68-77°F | 79-86°F | 50-59°F |
| Alternate hosts | Meadow rue (*Thalictrum* sp.) | Barberry | Not known |
| Annual recurrence | By uredospores from hills | By uredospores from hills | By uredospores from hills |
| Survival altitudes | About 1450 -2050 meter | About 1500 meter | About 1900 meter |

## Loose smut

## *Ustilago nuda tritici(Ustilago tritici)*

### Diagnostic Symptoms

- Infected plants is very difficult to detect in the field until heading.
- Infected heads emerge earlier than normal heads.
- Entire inflorescence is commonly affected and appears as a mass of olive-black spores, initially covered by a thin gray membrane.
- Once the membrane ruptures, the head appears powdery.
- Spores are dislodged, leaving only the rachis intact.
- In some cases remnants of glumes and awns may be present on the exposed rachis.
- Smutted heads are shorter than healthy heads due to a reduction in the length of the rachis and peduncle.
- While infected heads are shorter, the rest of the plant is slightly taller than healthy plants(Fig. 1).

### Etiology

### Systemic position

| | |
|---|---|
| Kingdom | Fungi |
| Division | Eumycota |
| Phylum | Basidiomycotina |
| Class | Teliomyetes |
| Order | Ustilaginales |
| Family | Tilletiaceae |
| Genus | *Ustilago* |
| Species | *Tritici* |

- Mycelium is hyaline primarily but turns to brown near maturity.It is septate, dikaryotic, and grows systemically inside the plant.
- Mycelia cells get transformed into brown, spherical, echinulate teleutospores.
- Teleutospores measure 5-9 μm in diameter, germinate readily producing promycelium or basidium consisting of four uninucleate cells.

- Basidium, unlike other smut causing fungi, do not produces basidiospores.
- Its cells germinates and give rise to uninucleate hyphae, the primary hyphae; a pair of sexually compatible uninucleate primary hyphae fuse and results in dikaryotic hypha oftcn called infection hypha.

**Disease Cycle**

- The pathogen perennates as dormant mycelium lying inside the kernel (seed), which has been infected by the infection hypha.
- Mycelium lies dormant within the seed until next season when they are sown.
- On seed germination and growth of the seedling, the mycelium becomes active, starts developing, and moves systemically along the growing seedling.
- Mycelium is hyaline during its growth through the plant, but it turns to brown at maturity.
- When the head emerges, the hyphae accumulate in the floral parts.Their cells are transformed into teleutospores, which completely fill in the spikelets.
- They first remain covered by a delicate silvery membrane, which soon bursts spreading almost all teleutospores to atmosphere.
- They are now wind blown leaving a bare rachis behind. This is usually the time when the healthy heads of adjacent plants are in flowering stage.
- Disseminated teleutospores lodge between the glumes and reach the feathery stigma germinate thereupon in moist stigmatic fluid giving out promycelium or basidium consisting of four cells.
- The basidium produces no basidiospores, but its cells germinate and produce short uninucleate, sexually compatible primary hyphae that fuse in pairs and give rise to dikaryotic mycelium the infection thread.
- Infection hypha penetrates the flower through the stigma or the young ovary wall and become established in the pericarp, integuments, and in the tissues of the embryo before the kernels become mature.
- Mycelium then becomes inactive and remains dormant, primarily in scutellum of the kernel.
- Fungal presence in no way happers the grain formation .
- When such infected kernels are sown in the next season the hyphae become active and show further course of their action.

## Favorable Conditions

- Frequent rain showers.
- High humidity.
- Cool temperatures (16-22$^0$C) during flowering season.

## Integrated Management

- Use of resistant varieties like NP710, 718,761,770, Bansipali 808, Bansi 224 etc.
- **Hot water treatment**: the seeds are soaked in water for 4-6 hours at 68-86$^0$F and then placed in water for 2 minutes at 120$^0$F, and finally taken out, dried and sown.
- **Solar energy treatment**: The seeds are soaked into water for four hours (8 am-12 noon), then taken out and dried in sun for four hours (12 noon-4 pm), and then sown.
- Treat the seed with Vitavax @ 2g/kg seed or Carbendazim 50% WP @ 2 g/ Kg seeds or Carboxin 75% WP @ 2 -2.5 g/Kg seeds or Carboxin 37.5% + Thiram 37.5% DS @ 3.0 g/Kg before sowing.
- In the standing crop, the plants showing yellowing of the boot leaf tip normally are the ones which will give smutted ear heads on emergence. Uproot such plants before ear emergence to reduce the infestation of healthy seeds at later stage.
- Use disease free seeds in the healthy field. For seed production, disease free field/areas to be identified for having crop without considerable inoculum load.
- Burry the infected ear heads in the soil, so that secondary spread is avoided.

## Karnal Bunt

***Neovosia indica***

## Diagnostic Symptoms

- Symptoms of karnal bunt are often difficult to distinguish in the field owing to the fact that incidence of infected kernels on a given head is low.
- Symptoms are most readily detected on seed after harvest.
- Black sorus, containing dusty spores is evident on part of the seed, commonly occurring along the groove.
- Heavily infected seed is fragile and the pericarp ruptures easily.

- Foul, fishy odor associated with karnal bunt.
- Odor is caused by the production of trimethylamine by the pathogen.
- Seed that is not extensively infected may germinate and produce healthy plants (Fig. 5).

## Etiology

## Scientific Classification

Kingdom : Fungi

Division : Eumoycota

Sub-Division : Basidiomyotina

Class : Teliomycetes

Order : Ustilaginales

Family : Tilletiaceae

Genus : *Neovosia*

Species : *indica*

- Teleutospore: spherical to oval with reticulation looking as curved spines on the episore, dark brown and measure 20-49 μ in diameter.
- Teleutospore wall consists of three layers; perisporium, episporium and endosporium.
- On germination, a teleutospore gives rise to a promycelium that bears long, sickle shaped sporidia in a cluster at its tip; each cluster or whorl generally consists of 60-185 sporidia called primary sporidia and are unicellular.
- Primary sporidia are incompatible with each other, so produce another crop of sporidia called secondary sporidia.
- The compatible primary sporidia and secondary sporidia fuse and develop into the dikaryotic hyphae or sporidia.
- The infection is caused by these dikaryotic hyphae or sporidia.

## Disease Cycle

- Disease is soil borne and pathogen perennates through teleutospores, which reach the soil either directly falling on the ground or through sticking on the surface of infected seeds and survive until flowering time of the next wheat crop.

- When favourable conditions become available at the time of flowering of the wheat crop, the perennating teleutospores germinate producing short, stout promycelia which give rise to primary sporidia in whorl.
- Primary sporidia produce secondary sporidia.
- Sporidia are wind or water splash disseminated to the flowers where upon the compatible primary and secondary sporidia fuse producing dikaryotic hyphae and or sporidia.
- These are the dikaryotic hyphae and or sporidia which cause primary infection of individual florets.
- Secondary spread of the pathogen within and between spikelets takes place through secondary sporidia and dikaryotic mycelia produced on the primary infected spikelets.

## Favorable Conditions

- Relative humidity over 70% favors teliospore development.
- Day time temperatures - 18-24 °C.
- Soil temperatures -17-21 °C.

## Integrated Management

- Grow resistant / tolerant varieties.
- Low-lying areas of the field accumulate water and are more prone to kernal bunt. Effective land leveling and drainage can reduce disease incidence.
- Decrease seed rate during sowing.
- Increase row spacing during sowing.
- Delayed sowing.
- Avoiding irrigation during the period of awn emergence and end of flowering may hinder disease development.
- Avoid lodging by using balance dose of Nitrogen and Potash.
- Plastic mulching or solarization reduces the chance of teliospore germination.
- Burn the stubble after harvesting.
- Crop rotation with non-host crop.
- Foliar spray with Propiconazole 25% EC @ 0.1 percent, if required repeat after 15 days.

## Powdery Mildew

### *Blumeria graminis* f.sp. *tritici*

### Diagnostic Symptoms

- Powdery mildew is characterized by white, cottony patches of mycelium and conidia on the surface of the plant.
- They can occur on all aerial parts of the plant including stems and heads, but are most conspicuous on the upper surfaces of lower leaves. The white colonies later turn dull gray-brown.
- Severe infections can cause stunting.
- Heads on the later, shorter tillers can become heavily diseased because they remain lower in the wheat canopy where humidity is high.
- As the growing season progresses, sexual fruiting structures known as cleistothecia appear as distinct brown-black dots within aging colonies on maturing plants.
- When severe, individual patches often merge and cover large areas of the stem, leaf surface, or head.
- Leaf tissue on the opposite side of the mildew pustules becomes yellow, later turning tan or brown.

### *Etiology*

Kingdom : Fungi

Division : Eumycota

Class : Leotiomycetes

Order : Erysiphales

Family : Erysiphaceae

Genus : *Blumeria*

Species : *graminis*

- Fungus produces septate, hyaline mycelium on leaf surface with short conidiophores.
- Conidia are single celled, elliptical, hyaline, thin walled and produced in chains.
- Dark globo se cleistothecia containing 9-30 asci develop with oblong, hyaline and thin walled ascospores.

## Disease Cycle

- Fungus remains in high hills during summers in infected plant debris as dormant mycelium and asci.
- Primary spread is by the asciospores.
- Secondary spread through airborne conidia.

## Favorable Conditions

- Temperature - 20-21°C.
- Prolonged cloudy weather

## Integrated Management

- Use resistant varieties if available.
- Use a correct and balanced fertilization program with proper levels of N, P and K.
- Monitor field regularly of first signs of the disease.
- Plan a crop rotation with non host plants.
- Eliminate volunteer plants and weeds from the field to disrupt its life cycle.
- Foliar spray withWettable Sulphur 0.2% or Carbendazim 0.1%, if require repeat after 15 days.

## Alternaria Blight

***Alternaria triticina***

## Diagnostic Symptoms

- Disease appears when wheat plants are 7-8 weeks old and becomes severe when the crop is mature.
- Infection is first evident as small, oval, discoloured lesions irregularly scattered on the leaves. As the lesions enlarge they become dark brown to grey and irregular in shape.
- Some are surrounded by a bright-yellow marginal zone. The lesions vary in size, reaching a diameter of 1 cm or more.
- As the disease progresses, several lesions coalesce to cover large areas, resulting in the death of the entire leaf.
- In some cases the leaf starts drying up from the tip, prematurely, when lesions appear.
- The lower most leaves are the first to show the signs of infection; the fungus gradually spreads to the upper leaves.

- In severe cases, similar symptoms are produced on the leaf sheath and stem, as well as the awns and glumes if spikes are infected at the pre-anthesis stage.
- If the spike is infected this early, seeds do not form.
- Infection at the dough stage of seed development results in glume infection, ear infection and seed infection.
- Heavily infected fields present a burnt appearance (Fig. 6).

## Etiology

### Scientific Classification

Kingdom : Fungi

Phylum : Ascomycota

Class : Dothideomycetes

Subclass : Pleosporomycetidae

Order : Pleosporales

Family : Pleosporaceae

Genus : *Alternaria*

Species : *triticina*

- Mycelium: initially hyaline, becoming olive-buff to deep olive-buff, branched, septate, 2-7 µm wide.
- Conidiophores: septate, unbranched or occasionally branched, erect, broader towards the distal end, on the host single or fasciculate, emerging through stomata, amphigenous, geniculate or straight, length variable, between septa, measure 17-28 × 3-6 µm.
- Conidia: acrogenous, borne singly or in chains of 2-4, smooth, irregularly ovoid, both ends rounded, or ellipsoid, or conical-ellipsoid, gradually tapering into a beak; beak concolorous with the main conidial body, straight, measure 20-37 × 3-7 µm.There are 1-10 transverse septa, 0 -5 longitudinal septa, constricted at septa, varying in size; length including beak 15-89 µm, width 7-30 µm

## Disease Cycle

- Pathogen perpetuates on infected seed.
- Primary spread is by externally seed-borne conidia.
- Secondary spread by air-borne conidia.

- The inoculum is mainly air borne, spread by means of conidia which infect the leaves to cause the blight.

## Favorable Conditions

- Temperature - 25°C
- High relative humidity.

## Integrated Management

- Use healthy seeds from certified sources.
- Practice the best hyegine measures.
- Infected crop should be destroyed by burning and ploughing.
- Apply adequate fertilizers.
- Delayed tillage of fields.
- Spray the crop with Mancozeb (0.2%) or Kresoxim-Mehtyl (0.1%). If required repeat after 15 days.

## Seed gall nematode

## Damage symptoms

- Infested seedlings show slight swelling or enlargement of the basal part of the stem.
- Leaves emerge from seedlings are crinkled or twisted , often folded with their tips held near the growing point .
- If severe infestation, seedlings may even die.
- Infested plants may show profuse tillering and may produce ear even 30-40 days earlier.
- Affeted ears are shorter.
- In glumes, seed is replaced by the gall which are smaller, darker colored and proportionally shorter.
- *A. tritici* when associated with Corynebacterium responsible for causing yellow ear rot or Tundu.
- Charaterized by production of light yellow slimy ooze on young leaves or ear head.
- Yellow slime, which can seen trickling down the tissue in humid weather, becomes, hard, brittle and brown on drying.
- Complete destruction of ear head.

## Systemic position

| | |
|---|---|
| Phylum | Nematoda |
| Class | Secernentea |
| Order | Tylenchida |
| Family | Anguinidae |
| Genus | *Anguina* |
| Species | *tritici* |

## Diagnostic Characteristics

### Female

- Adult female observed in the middle; 2.64-4.36 mm long; body coiled when heat relaxed; spear weak, 8-10 μm long with small knobs.
- Oesophagus with a lobed basal bulb; ovary single with one or two flexures anteriorly extending up to basal bulb, oocytes arranged in multiple rows about a rachis, valve posterior.
- Tail elongated conoid.

### Male

- Shorter than females, 2.04-2.4 mm long, more straight than females; bursa not covering the tail compleltely; spicules short and broad.

## Life Cycle

- Seed galls are dispersed along with seed during planting and harvest. In moist soil, seed galls release thousands of larvae.
- Wet weather favors larval movement and the infestation process.
- Nematode invades the crown and basal stem area, finally penetrating floral primordia. There they mature and produce large numbers of eggs.
- Seed galls develop in undifferentiated floral tissues.
- In the developing galls, the larvae mature into males and females, as the case may be. A single gall at this stage may contain 40 females and an equal number of males.
- They mate within the gall and the gravid females lay a large number of eggs.
- The young larvae on emerging from the eggs develop up to the second stage and then become dormant.
- They remain in that state in the dry galls till the next sowing season.
- There is only one generation in a year.

### Favourable conditions

- Loamy light soils.

### Integrated Management

- Use certified seed of resistance varieties only and clean seed by sieving or by using 2% salt water floatation to remove galls and prevent ear cockle diseases.
- Use the tolerant varieties such as C-306 for brown mite.
- Dry cleaning: Galls can be separated by coarse sieve from the healthy seeds.
- Winnowing or fanning.
- Brine floatation: 2% salt solution in place of plain water removes almost 100 % galls.
- Apply neem cake@ 80 Kg/acre or Carbofuran 3% CG @ 25 Kg/acre.

## Model Question Paper

### Section 'B'

### Long Answer Question

Each question carries 10 marks. Answer must be descriptive.

1. Briefly describe the diagnostic symptoms, causal organism, disease cycle and disease management of the following wheat disease

   (a) Brown or Leaf rust of wheat (b) Loose smut of wheat (c) Powdery mildew of wheat.

2. Describe in detail the symptoms, causal organism, disease-cycle and management of Karnal bunt of wheat.
3. Write the causal organism and major symptoms of any three of the following plant disease:

i) Yellow or Stripe rust of wheat (ii) Seed gall of wheat (iii) Alterneria leaf blight of wheat.

4. Describe the symptoms and disease management of any two plant diseases caused by the following pathogens:

   (i) *Ustilago tritici* (ii) *Puccinia graminis tritici* (iii) *Anguina tritici*

5. How do the three rusts of wheat differ? Describe in brief the different stages of any heteroecious rust occurring on wheat.
6. Describe the present views or recent work on annual recurrenc of wheat rusts in plants of India.

7. Describe the loose smut of wheat under heads, diagnostic symptoms, causal organism, disease cycle and disease management.
8. What are the important bunt diseases of wheat? Describe any one.
9. Distinguish between
   a. Localized and Systemic smut infection
   b. Monocyclic and polycyclic smut diseases
   c. *Tilletia caries* and *Tilletia foetida*
10. Write short notes on
   a. Disease cycle of loose smut
   b. Management of smut disease
   c. Loose smut of wheat

**Section 'B'**

**Short answer questions**

This question contains three questions. Short answers of not more than 100 words. Each question carries 5 marks.

1. Differentiate between yellow rust and brown rust of wheat.
2. How would you identify the following disease in the field : (a)Loose smut of wheat (b) Black or Stem rust of wheat
3. Give the disease cycle of powdery mildew of wheat.

**Section 'C'**

**Very short answer questions**

Note : Answer of each part should be 20-30 words. Each part contains 2 marks.

1. Name the pathogen or causal organism of following plant disease ; (a) Ear-cockle disease of wheat (b) Yellow or Stripe rust of wheat (c) Yellow ear rot of wheat (d) Powdery mildew of wheat.
2. Explain solar energy treatment of wheat seeds.
3. Give the different spore stages of any heteroecious rust.
4. Differentiate between Smut and Rust.
5. Name the three rust diseases of wheat with their causal organism.

**Section 'D'**

**Objective type questions**

Note :Each part carries 1 mark. Answer them accordingly.

A. Choose the correct answer from the following :

(i) Which of the following is an internally seed borne disease.

a. Powdery mildew of wheat  b. Brown of leaf rust of wheat

c. Loose smut of wheat  d. None of the Above

(ii) Karnal bunt of wheat is caused due to :

a. *Puccinia graminis tritici*  b. *Erysiphe graminis tritici*

c. *Tilletia indica*  d. *Clavibacter tritici*

(iii) Which of the following rust appears earliest on wheat ?

a. Brown or Leaf rust of wheat  b. Black or Stem rust of wheat

c. Yellow or Stripe rust of wheat  d. None of the Above

(iv) Who studies the detailed disease cycle of cereal rusts in India?

a. R.S. Singh  b. K.C. Mehta

c. B.B. Mundkur  d. E.J. Butler

(v) *Clavibacter tritici* is the causal organism of :

a. Flag smut of wheat  b. Black or Stem rust of wheat

c. Tundu or Yellow ear rot of wheat  d. Leaf blight of wheat

B. Fill in the blanks with suitable word (s) :

1. Loose smut of wheat is incited by ............
2. *Ustilago tritici* causes .................. disease.
3. Of the three rusts in wheat ............ requires lowest temperature for disease developement.
4. ..................... is known for his outstanding contribution on cereal rusts in India.
5. *Clavibacter tritici* is the causal organism of ..............
6. Smuts are obligate parasite, except ----------------------, which grows well in the soil.
7. The term bunt is a distortion of --------------, while smut is a German word --------------
8. Smut diseases caused by Tilletia are called---------------------
9. The fishy stink emitted in stinking smut of wheat is due to the chemical----------------------produced bt the fungus.
10. Loose smut of wheat and loose smut of --------------------are same except in host specificity.

C. State whether the following statements are *True* or *False* :

(i) Disease cycle of wheat rusts in India was described by K.C. Mehta.

(ii) Seed treatment with Captan is highly effective for the control of loose smut of wheat.

(iii) The rotten fish smell produced in the karnal bunt infected wheat fields is due to the production of trimethylamine by the causal fungus.

(iv) *Puccinia graminis tritici* is an autoecious rust fungus.

(v) Powdery mildew of wheat is caused by *Erysiphe graminis tritici.*

**Fig. 1:** Loose smut

**Fig. 2:** Stem rust

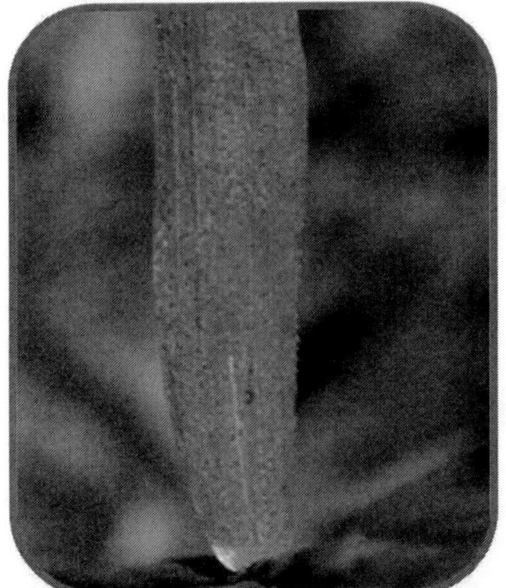

**Fig. 3:** Brown rust

**Fig. 4:** Yellow rust

**Fig. 5:** Karnal bunt

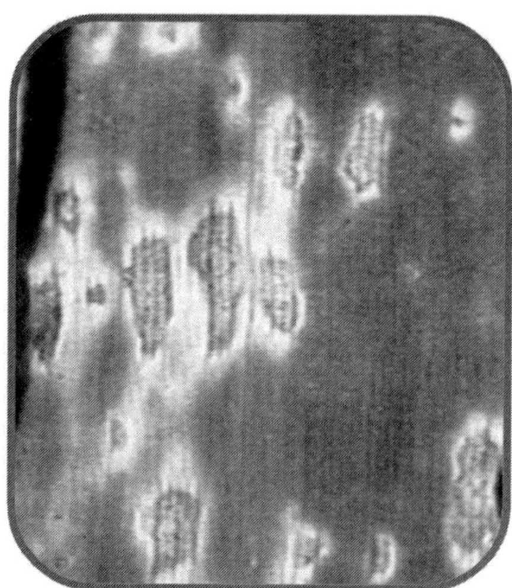

**Fig. 6:** Leaf blight

**Plate 1:** Photograph showing symptoms of major diseases of wheat *(See colour version on page 201)*

# 2

# Diseases of Sugarcane and Their Management

| Sl. No | Major Diseases | Causal Organism |
|---|---|---|
| 1 | Red rot | *Colletotrichum falcatum* |
| 2 | Smut | *Ustilago scitaminea* |
| 3 | Wilt | *Cephalosporium sacchari* |
| 4 | Grassy shoot | *Phytoplasma* |
| 5 | Ratoon stunting | *Clavibacterium xyli* sub sp. *xyli* |
| 6 | Pokkahboeng | *Fusarium moniliformae* and *Fusarium subglufinans* |
| 7 | Leaf scald | *Xanthomonas albilineans*(Ashby) |

## Red Rot

***Colletotrichum falcatum*(Perfect stage: *Physalospora tucumanensis*)**

### Diagnostic Symptoms

- Practically all above ground parts are affected, but stalks are the main targets of stack.
- Spindle leaves (third or fourth) starts drying.
- Further stalks becomes discolored and hollow.
- Diagnostic symptom is the formation of red streaks, interrupted by white patches , inside the stalk.This can be seen by splitting the stalks longitudinally.
- White patches are characteristics of disease, not found in other stalk rots.
- After splitting opened the diseased stalk, a sour smell due to conversion of glucose to alcohol.
- Red longitudinal lesions are also seen on the mid ribs of the leaves on which black color fruiting body (Acervuli) are formed (Fig.1).

## Etiology

| | |
|---|---|
| Phylum | Deuteromycotina |
| Class | Hyphomycetes |
| Order | Melanconiales |
| Family | Melanconiaceae |
| Genus | *Colletotrichum* |
| Species | *falcaturm* |

- Hyphae: profusely branched septate, hyphae containing oil droplets.
- Acervuli:black, minute velvetty acervuli with long, rigid bristle-like, septate setae.
- Conidiophores: a closely packed inside the acervulus, which are short, hyaline and single celled.
- Conidia: single celled, hyaline, falcate, granular and guttulate.
- Perithecia: dark brown to black with a papillate ostiole.
- Asci : clavate, unitunicate and eight-spored.
- Ascospore: hyaline, straight or slightly curved, ellipsoid or fusoid and unicellular which measure 18-22 μm x 7-8μm.

## Disease Cycle

- Pathogen can survive as chlaymydospores in crop debris or in soil.
- Pathogen produces acervuli and conidia in the soil that serve as primary inoculums and infect the sets. Diseased sets also give rise to infected plants.
- Secondary infection that spreads the disease is brought about by conidia formed on the mid rib lesions of the leaves.
- Dissemination of conidia is through wind , rain , heavy dew and irrigation waters.

## Favorable Conditions

- Successive ratoon cropping.
- Monoculturing of sugarcane.
- Water logged conditions and injuries caused by insects.
- Low temperature.

## Integrated Management

- Grow resistant and moderately resistant varieties CO 62198, CO 7704 , CO 8001, CO8201.

- Select the setts from the disease free fields or disease free areas.
- Setts can be treated with by moist hot air at 54°C for 2 hours and by aerated steam at 52 °C for 4 to 5 hours.
- Soak the setts in Carbendazim (0.1%) or Triademcfon (0.05%) solution for 15 minutes before planting.
- Adopt crop rotation by including rice and green manure crops.
- Aviod ratooning of the diseased crop.

## Smut

### *Ustilago scitaminea*

### Diagnostic Symptoms

- Production of long whip-like structure from the gwoing point of of affected cane.
- Whip may be few millimeters to 20 mm in diameter.
- It is unbranched and made up of a fairly hard core of parenchyma and fibrovascular element in which the fungal hyphae ramify with the telioospore.
- Whip enclosed at first in a thin silvery sheath enclosing mass of black powdery spores.
- Initally thin canes with elongated internodes later become reduced in length.
- Profuse sporuting of lateral buds with narrow erect leaves especially in ratton crop( Fig.2).

### Etiology

| | |
|---|---|
| Phylum | Basidiomycotina |
| Class | Teliomyetes |
| Order | Ustilaginales |
| Family | Tilletiaceae |
| Genus | *Ustilago* |
| Species | *scitaminea* |

- Pathogen is true culmicolous (stem infectiong) smut.
- Fungal hyphae are primarily intercellular and collect as a dense mass between the vascular bundles of host cell and produce tiny black spores.
- Teliospores are spherical light brown, and 10.5 to 17.04 in diameter.

- Teliospores germinate to produce 3-4 celled, hyaline promycelium and produce 4 sporidia (basidiospore) which are hyaline and oval shaped with pointed ends.
- Dikaryotic phase is established through fusion between sporidia, between cells of promycelium or between cells of hyphae developing from the germinating teliospores.
- Dikaryotic infection hypha invades the host tissue.
- When the whip is fully emerged the membrane ruptures releasing teliospores in the air.

## Disease Cycle

- Pathogen survives as smut spores and dormant mycelium present in or on the infected setts and in the infected plant materials in the soil.
- Primary spread of the disease is through diseased seed-pieces (setts).
- Secondary spread in the field is mainly through the wind borne smut spores.
- Disease spread centrifugally in the field.

## Favorable Conditions

- Continuous ratooning.
- Monoculture.
- Dry weather.

## Integrated Management

- Plant healthy setts taken from disease free area.
- Grow resistant varieties like Co 7704, COC 85061 and COC 8201.
- Physical or chemical treatment of sets before planting.
- Physical method include hot water treatment, the sets are dipped in water at 55$^0$C to 60 $^0$C for 10 minute, or exposure of sets to moist hot air at 54$^0$C for eight hours .
- Hot water plus Bayleton (0.1%) at 50$^0$C completely eliminate the smut from sets.
- Discourage ratooning of the diseased crops having more than 10 per cent infection.
- Remove and destory the smutted clump (collect the whips in a thick cloth bag/polythene bag and immerse in boiling water for 1 hr to kill the spores).
- Follow crop rotation with green manure crops or dry fallowing.

## Wilt

*Cephalosporium sacchari*

### Diagnostic Symptoms

- Yellowing and withering of crown leaves.
- Shrinkage and withering of canes.
- Leaves dry up and stem develop hollowness in the core.
- Core shows the reddish discolouration with longitudinal red streaks passing from one internode to another.
- Spindle shaped cavities tapering towards the nodes develop in each internode( Fig.6).

### Etiology

| | |
|---|---|
| Kingdom | Fungi |
| Division | Eumycota |
| Phylum | Duteromycotina |
| Class | Hyphomycetes |
| Order | Tuberculariales |
| Family | Tuberculariaceae |
| Genus | *Cephalosporium* |
| Species | *sacchari* |

- Mycelium: hyaline, septate and thin walled, produced many conidiophores as lateral branches.
- Conidiophores: simple or branched and produce single celled, oval to elliptical, hyaline, microconidia.
- Microconidia: hyaline, ovoid, or oblong-ellipsoidal, aseptate, measure 4-12x1-3 μm.
- Macroconida: Not produced.This is the feature which differentiates the fungus from that of Fusarium.

### Disease Cycle

- Pathogen is soil-borne and remains survives in the soil as saprophyte for 2-3 years.
- Primary infection occurs by sowing infected seed pieces.
- Secondary spread is by wind, rain and irrigation water.

## Favorable Conditions

- High temperature -30-35°C.
- Low humidity -50-60%.
- Excess doses of nitrogenous fertilizers.
- Low soil moisture.
- Alkaline soils.

## Integrated Managemet

- Select the seed material from the disease-free plots.
- Dip the setts in 40 ppm Boran or Manganese for 10 minutes or in Carbendazim (0.05%) for 15 minutes.
- Amend soil with boric acid (15 kg/ha) and sett treatment with Carbendazim (0.1 %) .
- Crop rotation with non host crops.
- Avoid excess of fertilizers in field, especially nitrogen.
- Avoid the practice of ratooning in diseased fields.
- Burn the trashes and stubbles in the field.

## Grassy Shoot

***Phytoplasma***

## Diagnostic Symptoms

- Characterized by proliferation of vegetative buds from the base of the cane giving rise to crowded bunch of tillers bearing narrow leaves.
- Leaves become pale yellow to completely chlorotic, thin and narrow.
- Plants appear bushy and 'grass-like' due to reduction in the length of internodes premature and continuous tillering.
- Cane formation rarely occurs in the affected clumps, if formed, thin with shorter internodes having aerial roots at the lower nodes ( Fig.3).

## Etiology

- Two types of bodies are seen in ultra thin sections of phloem cells of infected plants.
- Spherical bodies of 300-400 nm diameter and filamentous bodies of 30-53 mm diameter in size.

## Disease Cycle

- Pathogen is perpetuated through crop rattoning.
- Primary transmission isthrough infected seed material.
- Secondary spread is by aphid,grass hopper and dodder.

## Integrated Management

- Select setts from diseased free area.
- Grow resistant varieties *viz.*, Co 86249, CoG 93076 and Coc 22.
- Pre-treating the healthy setts with hot water at 52°C for 1 hour before planting.
- Complete cure of this disease was achieved by treating single-bud setts with 500 or 1000 ppm of Ledermycin or 250 ppm of Achromycin, Terrarriycin and Erythromycin.
- Rogue out infected plants in the secondary and commercial seed nursery.
- Spray Dimethoate @ 0.1 % to control insect vector.
- Avoid ratooning if GSD incidence is more than 15 % in the plant crop.

## Ratoon Stunting

*Clavibacter xyli* sub sp. *xyli* (*Rickettsia Like Organism - RLO)*

## Diagnostic Symptoms

- It is called "ratoon" stunting because the disease is more severe on ratton crops than on 'plant cane' crops.
- It causes 5-15% loss in yield without the grower even noticing the disease because there no recognizable symptoms, except stunting, which too is not distinctive.
- Stunted growth, reduced tillering, thin stalks with shortened internodes and yellowish foliage.
- Show signs of wilting during summer.
- On spilliting open the cane longitudinally, two types of discoloration is seen in the pith,Orange-red vascular bundles in shades of yellow at the nodes are seen in the infected mature canes.
- Pink color is seen near the nodes in young canes (Fig.4).

## Etiology

Kingdom Bacteria

Division Actinobacteria

Order Actinomycetales

Family Microbacteriaceae

Genus *Clavibactor*

Species *xyli*

- Slow growing, gram positive, rod shaped.
- Pathogen is present in the xylem cells of infected plants so called as fastidious xylem colonizing bacterium.
- They are small, thin, rod shaped or coryneform, measuring 0.3-0.5x1-3 microns , with length of the filamentous being 10 microns or more.
- Only sap transmitted and no insect vector is known.
- Disease spreads through cutting knives and seed setts from diseased canes.

**Disease Cycle**

- Disease is primarily transmitted through seed-pieces used for planting.
- Harvesting implements contaminated with infected juice also help spread the pathogen.
- Johnson grass, maize and elephant grass act as carriers of the RSD bacterium and may serve as the source of noculums.

**Favorable Conditions**

- Optimum temperature -26-30°C.
- Moisture stress.

**Integrated Management**

- Select the setts from disease free fields.
- Treating the setts in hot water at 50°C for about 2 hours.
- Remove and burn the clumps showing the disease incidence.
- No noculums of weak or stunted crops.

**Pokkahboeng**

*Fusarium moniliforme* and *Fusarium subglufinans*

**Diagnostic symptoms**

- General symptoms of Pokkahboeng are mainly of three types
- Chlorotic Phase:

  - Chlorotic condition towards the base of the young leaves and occasionally on the other parts of the leaf blades.
  - Frequently, a pronounced wrinkling, twisting and shortening of the leaves accompanied the malformation or distortion of the young leaves.
  - Base of the affected leaves is seen often narrower than that of the normal leaves.
- Acute Phase or Top-Rot Phase:
  - Most serious stage of Pokkahboeng is a top rot phase. The young spindles are killed and the entire top dies.
  - Leaf infection sometimes continued to downward and penetrates in the stalk by way of a growing point.
  - In advanced stage of infection, the entire base of the spindle and even growing point showed a malformation of leaves, pronounced wrinkling, twisting and rotting of spindle leaves.
  - Red specks and stripes also developed.
- Knife-cut Phase :
  - Characterized by one or two or even more transverse cuts in the rind of the stalk /stem in such a uniform manner as if, the tissues are removed with a sharp knife.
- This is an exaggerated stage of a typical ladder lesion of a Pokkahboeng disease ( Fig.5).

## Disease Cycle

- Air-borne disease.
- Primarily transmitted through the air-currents.
- Secondary transmission is through the infected setts, irrigation water, splashed rains and soil.

## Favourable conditions

- Temperature- 20-30°C.
- Relative humidity ->70 to 80%.
- Cloudy weather, drizzling rains.

## Integrated Management

- Plant healthy seed material.

- Use resistant varieties.
- Foliar spray with Carbendazim (0.1%) or Copper oxychloride(0.3%), Mancozeb (0.3%), if required repeat after 15 days.
- Canes showing 'top rot' or 'knife cut' should be rouged out.

## MODEL QUESTION PAPER

### Section 'A'

### Long answer questions

1. Describe in detail the symptoms, etiology, disease cycle and management of red rot of sugarcane.
2. (a) Briefly describe the most distinguishing symptoms, causal organism and management of smut of sugarcane.

   (b) Briefly describe the most distinguishing symptoms, causal organism and management wilt of sugarcane.
3. Describe in detail the most distinguishing symptoms, causal organism, etiology, disease cycle and management of grassy shoot disease of sugarcane.
4. Describe in brief the most distinguishing symptoms, transmission and disease management of ratoon stunting.
5. Describe the disease Pokkahboeng giving symptoms, pathogen, disease cycle and management.

### Section 'B'

### Short answer questions

6. Write the causal organism and major symptoms of the following disease:

   (i) Smut of sugarcane  (ii) Wilt of sugarcane
7. Give the integrated disease management of wilt of sugarcane.
8. Explain the symptom and management of smut of sugarcane.

### Section 'C'

### Very short answer questions

9. (a) Name the pathogen causing the following plant disease :

   (i) Red rot of sugarcane  (ii) Pokkahboeng

   (iii) Wilt of sugarcane  (iv) Ratoon stunting of sugarcane

(b) Name the disease caused by the following pathogens :

(i) *Cephalosporium sacchari*

(ii) *Xanthomonas albilineans*(Ashby)

(iii) *Colletotrichum falcatum*

(iv) *Clavibacter xyli* sub sp. *xyli*

(c). Write down the Integrated management schedule of whip smut of sugarcane ?

(d) What is grassy shoot ?

**Section 'D'**

**Objective type questions**

10. Choose the correct answer from the following :

(i) Grassy shoot disease is caused by :

a. Phytoplasma b. Vioroid

c. Prions d. Virus

(ii) *Thanatephorus cucumaris* is the perfect state of :

a. *Colletotrichum falcatum* b. *Rhizoctonia solani*

c. *Rhizoctonia bataticola* d. *Fusarium oxysporum* f. sp. *vasinfectum*

(iii) *Ustilago scitaminea* causes :

a. Smut of sugarcane b. Wilt of cotton

c. Damping off of tobacco d. Wilt of sugarcane

(iv) Red rot is a destructive disease of :

a. Cotton b. Tobacco

c. Sugarcane d. Potato

(v) The perfect state or teleomorph of *Colletotrichum falcatum* is :

a. *Physalsospora tucumanensis* b. *Glomerella tucumanensis*

c. *Thanatephorus cucumeris* d. *Fusarium oxysporum*

**Fig. 1:** Red rot

**Fig. 2:** Smut

**Fig. 3:** Grassy shoot

**Fig. 4:** Ratton stuting

**Fig. 5:** Pokkahboeng

**Fig. 6:** Wilt

**Plate-2.** Photograph showing symptoms of major diseases of Sugarcane *(See colour version on page 202)*

# 3

# Diseases of Sunflower and Their Management

| Sl. No | Major Diseases | Causal Organism |
|---|---|---|
| 1 | Root rot or charcoal rot | *Macrophomina phaseolina* (Tassi) Goid |
| 2 | Alternaria blight | *Alternaria helianthi* (Hansf.) Tubaki& Nishi |
| 3 | Sclerotinia wilt and stem rot | *Sclerotinia sclerotiorum* (Lib.) de Bary |
| 4 | Rust | *Puccinia helianthi* Schwein |
| 5 | Downy mildew | *Plasmopara halstedii* Farl. Berl. & De Toni |
| 6 | Necrosis | Tobacco streak virus |
| 7 | Head rot | *Rhizopus arrhizus* Fischer |

## Alternaria Blight

### *Alternaria helianthi*

### Diagnostic Symptom

- Circular to oval, dark brown to black spots appear on the leaves.The spots enlarge into round to irregular spots with concentric rings and coalesce causing blighting and withering of leaves.
- These are surrounded by chlorotic zone with grey white necrotic center.
- Spot appear first on the lower leaves and as the plant grows the spots subsequently appears on middle and upper leaves.
- Stem lesions begin as dark flecks which enlarge to form long, narrow lesions which may also coalesce to a larger blackened area resulting in stem breakage.
- In severe cases, the lesions appear on petioles, ray florets and head as brown round spots about 1cm diameter with a slight depression in the centre.
- Sometimes rotting of flower heads also occur (Fig.1).

### Etiology

Phylum Duteromycotina

Class Hyphomycetes

| | |
|---|---|
| Order | Hyphomycelates |
| Family | Dematiacaeae |
| Genus | *Alternaria* |
| Species | *helianthi* |

- Facultative parasite.
- Pathogen produces cylindrical conidiophores and conidia.
- Conidiophores are pale grey-yellow coloured, straight or curved, geniculate, simple or branched, septate and bear single conidium.
- Conidia are cylindrical to long ellipsoid, straight or slightly curved, 1 to 2 septate with longitudinal septa and pale grey-yellow to pale brown and size are 25-80 (120) x 8-11 microns.

**Disease cycle**

- Pathogen overwinters as mycelium on infected plant residues and in dry conditions survives for 20 weeks in so.
- Primary source of noculums are spores present in the soil and infected seed.
- Secondary spread is mainly through windblown conidia.

**Favourable Conditions**

- Air temperatures – 25 to 30$^{o}$C
- Rainy weather.
- Late sown crops are highly susceptible.

**Integrated Management**

- Field sanitation.
- Deep summer ploughing.
- Use of resistant or tolerant variety like B.S.H.1 .
- Seed treatment with Thiram or Carbendazim at 2 g/kg.
- Spray Mancozeb ( 0.2%) or Copper oxychloride (0.3%) Difenoconazole (0.1%) or Chlorothalonil (0.2%), if required repet after 15 days.
- Proper spacing.
- Application of well rotten manures.
- Practice crop rotation.
- Planting in mid-september.
- Remove and destroy the diseased plants.

## Sclerotinia wilt and rot

### *Sclerotinia sclerotiorum*

### Diagnostic Symptoms

- Early symptoms of the disease are noticed forty days after sowing.
- Stem is usually infected at or near the soil line.
- Brownish lesion develops at the base of the stem and eventually girdles the plant. A white fan-like growth forms over the infected tissues and often radiates over the soil surface.
- Sickly appearance of plants can be spotted from a distance and a row effect can be observed in heavily infected soil. Later the entire plant withers and dies.
- The lesion grows up the stem, destroying the cortical tissue and leaving the fibrous vascular strands as the tissues dry out.
- White cottony mycelium and mustard seed type sclerotial bodies are conspicuous on the affected stem near soil level. This also causes collar rot in the seedling stage ( Fig.2).

### Etiology

### Scientific Classification

Kingdom: Fungi

Division : Ascomycota

Class : Leotiomycetes

Order : Helotiales

Family : Sclerotiniaceae

Genus : *Sclerotinia*

Species : *sclerotiorum*

- Pathogen form hard blackish sclerotinia which germinate and produce cup shaped brown color apothecia.
- Sclerotia borne superficially usually on dense white mycelium, globose to cylindrical, but quite variable in shape, 2-15x2-15µm with black outer rind and white inner contex.
- Apothecia arising one to several form a sclerotium ochraceous often darker at the base of the stipe, receptacle 2-8 mm broad applanate to slightly concave, often with control depression frequently with an undulate margin, tapering to form a stipe 3-10 mm long, 1-2 mm wide.

### Disease Cycle

- Pathogen survives between crops in the soil as sclerotia.
- Primary infection occurs when soilborne sclerotia germinate.
- Germinating sclerotia form small, tan, cup-shaped structures called apothecia, which release millions of airborne ascospores (secondary infection) that colonize dead plant parts and infect host tissues.

### Favorable Conditions

- Cool wet weather.
- Optimum temperature-15$^0$C
- Leaf wetness –Approx. 48-72h
- Dense plant stands.

### Integrated Management

- Deep ploughing during summer.
- Crop rotation of 3-4 years.
- Timely sowing should be done.
- Use resistant/ tolerant varieties.
- Soil amendment with FYM @ 5 tonnes/acre.
- Avoid moisture stress during high summer and water logging conditions.
- Seed treatment with Trichoderma @ 6g/kg seed.
- Addition of Trichoderma in soil at 10g/kg and soil amendments like castor cake, neem cake, oat straw reduces disease incidence

## MODEL QUESTION PAPER

### Section 'A'

### Long answer questions

1. Describe in detail the symptoms, etiology, disease cycle and management of Alternaria blight sunflower.
2. Briefly describe the most distinguishing symptoms, causal organism, disease cycle and disease management of Sclerotinia stem rot disease of sunflower.
3. Briefly describe the most distinguishing symptoms, etiology and disease management of any of the following plant disease :

   (i) Alternaria blight of sunflower

   (ii) Sclerotinia stem rot diseases of sunflower.

4. Describe white blisters and Alternaria blight of sunflower under the following headings.
   (i) Etiology
   (ii) Symptoms
   (ii) Favorable symptoms
   (iv) Disease management
5. Describe in detail the symptoms, causal organism, disease cycle and management of rust of sunflower.

**Section 'B'**

**Short answer questions**

1. Describe the chemical management of Alternaria blight of sunflower.
2. Write etiology and disease cycle of Alternaria blight of sunflower.
3. Describe the diagnostic symptoms of Sclerotinia stem rot of sunflower.

**Section 'C'**

**Very short answer questions**

1. Name the pathogen causing the following plant diseases:
   (i) Rust of sunflower
   (ii) Alternaria blight of sunflower
   (iii) Sclerotinia stem rot of sunflower
   (iv) Downy mildew of sunflower
2. Give the name of different types of spore stages found in *Puccinia helianthi*
3. Name the disease caused by the following pathogens:
   (i) *Macrophomina phaseolina*
   (ii) *Alternaria helianthi*
   (iii) *Sclerotinia sclerotiorum*
   (iv) *Plasmopara halstedii*
4. Suggest at least four resistant varieties of sunflower against Alternaria blight disease of sunflower.

## Section 'D'

## Objective type questions

Note: Choose the correct answer from the following :

(i) Which one of the following diseases is caused by *Rhizopus arrhizus ?*

a. Head rot of sunflower
b. Downy mildew of sunflower
c. Sunflower necrosis
d. None of the above

(ii) Which one of the following pathogens cause leaf blight of sunflower?

a. *Puccinia helianthi*
b. *Alternaria helianthi*
c. *Cercosporidium personatum*
d. *Melampsora lini*

(iii) Which pathogen form hard blackish sclerotinia which germinate and produce cup shaped brown color apothecia.

a. *Puccinia helianthi*
b. *Alternaria helianthi*
c. *Sclerotinia sclerotiorum*
d. *Melampsora lini*

(iv) Which one of the pathogens causes sunflower necrosis?

a. Tobacco streak virus
b. *Puccinia helianthi*
c. *Sclerotinia sclerotiorum*
d. *Mycosphaerella spp.*

(v) Fungicides used for the management of Alternaria blight of sunflower ?

a. Mancozeb
b. Copper oxychloride
c. Difenoconazole
d. All of the above

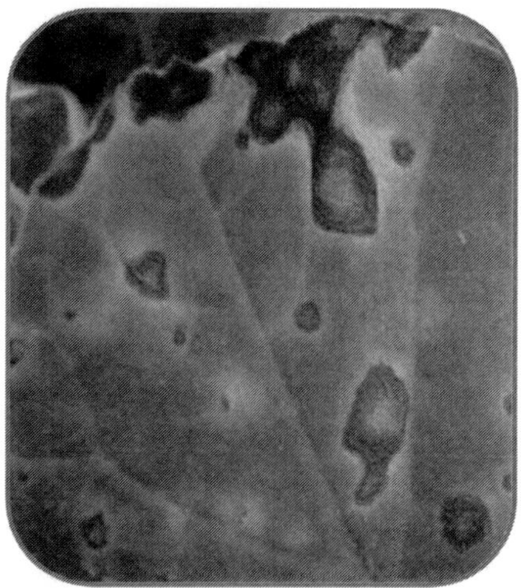

**Fig. 1:**.Alternaria leaf blight

**Fig. 2:** Sclerotinia rot

**Plate 3:** Photograph showing symptoms of major diseases of Sunflower *(See colour version on page 203)*

# 4

# Diseases of Mustard and Their Management

| Sl.No | Major Diseases | Causal Organism |
|---|---|---|
| 1 | White rust | *Albugo candida* |
| 2 | Alternaria blight | *Alternaria brassicae*, *A. brassicicola*, and *A.raphani* |
| 3 | Downy mildew | *Peronospora parasitica* |
| 4 | Powdery mildew | *Erysiphe cruciferarum* |
| 5 | Sclerotinia stem rot | *Sclerotinia sclerotiorum* |
| 6 | Club root | *Plasmodiophora brassicae* |
| 7 | Bacterial blight/ Black rot | *Xanthomonas campestris pv. campestris* |

## White Rust

### *Albugo candida*

### Diagnostic Symptoms

- Intially chlorotic (yellowed) lesions and sometimes galls appeared on the upper leaf surface.
- There are corresponding white blister-like dispersal pustules of sporangia on the underside of the leaf.
- Branches and flower parts deformed and pathogen stimulates hypertrophy and hyperplasia.
- Flower and young fruits may also malformed and pod lack seed development.
- Disease is sometimes called the cancer of rapeseed as one phase of this disease shows grotesque malformation of the young shoots and inflorescence (Fig.1).

### Etiology

### Scientific Classification

Phylum : Heterokontophyta

Class : Oomycota

Order : Peronosporales

Family : Albuginaceae

Genus : *Albugo*

Species : *candida*

- Pathogen is obligate parasite and it needs a living host to grow and reproduce.
- Mycelium: Intercellular producing small globose, knob shaped haustoria in the host cells.
- Sporangiophores: hyaline, clavate, free from each other laterally and are very thick walled especially towards base.
- Sporangia: hyaline, nearly oval to spherical with uniform thin walled and 14-16 x16-20µm in size.It may germinate by means of germ tube or through zoospores.
- Zoospores: Biflagellate, lose their flagella, come to rest, and then germinate by germtube which penetrates the epidermis of the host directly or through the stomata.
- Oogonium: globose, terminal or intercalary and multinucleate.
- Antheridium :clavate, pargynous and contains 6-12 nuclei.
- Oospores : globose, chocolate brown,30-35µm in diameter. It germinate by formation of vesicles in which zoospores are formed or the oospores may germinate by production of one or two simple or branched germ tube by release of zoospores from sessile vesicles.
- Germination of sessile vesicles is most common.

## Disease Cycle

- Pathogen reproduce by producing both sexual spores (called oospores) and asexual spores (called sporangia) in a many-stage (polycyclic) disease cycle.
- Thick-walled oospores are the main overwintering structures, but the mycelium can also survive in conditions where all the plant material is not destroyed during the winter.
- In the spring, the oospores germinate and produce sporangia (sporangium) on short stalks called sporangiophores that become so tightly packed within the leaf that they rupture the epidermis and are consequently spread by the wind.
- Liberated sporangia can either germinate directly with a germ tube or begin to produce biflagellate motile zoospores.

- Zoospores then swim in a film of water to a suitable site and each one produces a germ tube – like that of the sporangium – that penetrates the stoma.
- When the oomycete has successfully invaded the host plant, it grows and continues to reproduce.

## Favorable Conditions

- Cool and moist condition.
- Optimal temperature – 13 – 25 °C.
- Light rain or irrigation.
- Leaf surfaces need to remain wet for at least 2 to 3 hours .

## Integrated Management

- Sanitation.
- Use of resistant cultivars.
- Early sowing preferably in the first fortnight of October.
- Use the seed from stag-head free plants to avoid carry over of oospore through seeds.
- Treat the seed with Apron 35 SD @ 6gm/kg seed.
- Foliar spray the crop with Mancozeb + Carbendazim (0.2%) at the onset of the disease, repeat the spray after 15 days interval.
- Destroy the diseased plant debris.
- Crop rotation.

## Alternaria blight

***Alternaria brassicae*, *A. brassicicola*, *and* *A.raphani*.**

## Diagnostic Symptoms

- Symptom first appear on the lower leaves as small circular brown necrotic spots which slowly increase in size.
- Many concentric spots coalesce to cover large patches showing blightening and defoliation in severe cases.
- Circular to linear, dark brown lesions also develop on stems and pods, which are elongated at later stage.
- Infected pods produce small, discolored and shriveled seeds( Fig.2).

## Etiology

### Scientific Classification

Kingdom : Fungi

Phylum : Ascomycota

Class : Dothideomycetes

Subclass : Pleosporomycetidae

Order : Pleosporales

Family : Pleosporaceae

Genus : *Alternaria*

Species : *brassicae/brassicicola*

### *Alternaria brassicae*

- Mycelium: Immersed, hyphae branched, septate, hyaline, smooth,4-8µm thick.
- Conidiophores: arising in groups of 2-10 or more from the hyphae which are straight or curved.
- Conidia: light olive coloured with transverse and longitudinal septa. These are around 3-5 septate and conidia are borne in chain over short inoculums.

### *Alternaria brassicicola*

- Mycelium: immersed, hyphae branched, septate, hyaline, olivaceous brown, smooth, upto 70 µm long and 4-8µm thick.
- Conidiophores: arising in groups of 2-12, emerging through stomata, usually simple, erect, straight or curved, smooth upto 70µm long, 5-8 µm thick.
- Conidia: mostly in chains of upto 20 or more, sometimes branched, straight, nearly cylindrical, usually slightly tapering towards the apex, or obclavate, the basal cell rounded, the beak usually almost nonexistent, are light olive coloured with1-11 mostly less than 6 transverse septa.

### Disease Cycle

- Pathogens survive between mustard and canola crops in infested crop debris and on seed.
- These pathogens are disseminated as conidia within and among fields by splashing water and wind

- Pathogens infect their hosts by direct penetration or through wounds and natural openings.

## Favorable Conditions

- Warm and humid weather.
- Relative humidity-> 70 percent.
- Optimum temperature-12-20$^0$C.
- Intermittent rain.

## Integrated Management

- Plant high quality seed free from the Alternaria black spot pathogens.
- Practice a three year or longer rotation to non-hosts such as small grains.
- Promptly incorporate crop residues after harvest to hasten their breakdown.
- Eliminate volunteer canola, mustard and cruciferous weeds.
- Early sowing in first fortnight of October .
- Spray the crop with Mancozeb (0.2%) or Iprodione (0.2%), if required repeat after 15 days.

## Downy Mildew

*Peronospora parasitica*

## Diagnostic Symptoms

- Initially grayish white irregular necrotic patches develop on the lower surface of leaves.
- Later under favorable conditions, brownish white fungal growth may also be seen on the spots.
- Most conspicuous and pronounced symptom is the infection of inflorescence causing hypertrophy of the peduncle of inflorescence and develop stag head structure ( Fig.3).

## Etiology

## Scientific Classification

Kingdom : Fungi

Division : Eumycota

Class : Oomycetes

Order : Peronosporales

Family : Peronosporaceae

Genus : *Peronospora*

Species : *parasitica*

- Obligate parasite
- Mycelium: strictly intercellular, hyaline with large, finger shaped haustoria.
- Sporangiophores: 100-300µm long and unbranched for a major portion of their length.Dichotomous branching occurs, 6-8 times, at the tip. They are at acute angle with each other .A single sporangium is borne at the tip of each ultimate branch.
- Sporangia: dispersed by rain splash or wind and may germinate by means of germ tube.
- Oogonia: pale yellow, irregularly round, and swollen into crest like folds.
- Antheridia: tendril like and are produced on sepatate hyphae.
- A fertilization tube grows from the antheridium through the receptive papilla towards a central body in the ooplasm to discharge a single male nucleus.The two nuclei fuse and initiate the uninucleate oospore.
- Oospores: globose, measuring 26-43µm in diameter and germinate by germ tube .

## Disease Cycle

- Pathogen survives between crops in the soil as oospores.
- Infection occurs when soilborne resting structures called oospores germinate and produce sporangia under moist, cool conditions.
- Sporangia are produced on the underside of leaves in the evening and are released during the day as leaves dry.
- Windblown sporangia land on leaves and directly penetrate leaves and flowers.

## Favorable Conditions

- Cool, moist conditions.
- Optimum temperature-15-20°C
- Relative humidity-70 percent.

## Integrated Management

- Plant resistant or tolerant varieties.

- Seed treatment with Apron SD-70 (2g/kg seed)
- Sow the crop on normal planting date.
- Destroy diseased plant parts.
- Foliar spray with Mancozeb + Metalaxyl (0.2%), if needed repeat after 15 days of first spray.

## Sclerotinia Stem Rot

*Sclerotinia sclerotiorum.*

## Diagnostic Symptoms

- Initial symptoms appear as the plant canopy closes over rows during flowering and pod development.
- Lodged stems in contact with the soil develop watery lesions, with snowy white mycelium and black, irregularly shaped sclerotia become apparent as disease progresses.
- When the stem is completely girdled by such lesions, the plants wilt and die.
- When the crop is at seed stage, the plants tend to lodge, touching the siliquae to the soil level.
- Such plant show rotting of the siliquae with fungal growth along with the sclerotial bodies just above the soil level( Fig.4).

## Etiology

## Scientific Classification

Kingdom : Fungi

Division : Ascomycota

Class : Leotiomycetes

Order : Helotiales

Family : Sclerotiniaceae

Genus : *Sclerotinia*

Species : *sclerotiorum*

- Pathogen is characterized by the formation of hard blackish sclerotinia which germinate and produce cup shaped brown color apothecia.
- Sclerotia borne superficially usually on dense white mycelium, globose to cylindrical, but quite variable in shape, 2-15x2-15µm with black outer rind and white inner contex.

- Apothecia arising one to several form a sclerotium ochraceous often darker at the base of the stipe, receptacle 2-8 mm broad applanate to slightly concave, often with control depression frequently with an undulate margin, tapering to form a stipe 3-10 mm long, 1-2 mm wide.

## Disease Cycle

- Pathogen survives between crops in the soil as sclerotia.
- Infection occurs when soilborne sclerotia (dormant resting structures) germinate when the soil surface is continuously wet for at least two weeks.
- Germinating sclerotia form small, tan, cup-shaped structures called apothecia, which release millions of airborne ascospores that colonize dead plant parts (such as senescent flowers) and infect host tissues.
- Sclerotia can survive in soil for up 8 to 10 years.

## Favorable Conditions

- Cool wet weather.
- Optimum temperature for apothecia formation-15$^0$C.
- Continuous leaf wetness is required for infection by ascospores-App. 48-72h
- Dense plant stands.

## Integrated Management

- Plant high quality seed free from sclerotia.
- Practice a four-year or longer rotation to non-hosts such as corn or small grains.
- Avoid excessive irrigation and fertilization.

## MODEL QUESTION PAPER

### Section 'A'

### Long answer questions

1. Describe in detail the symptoms, etiology, disease cycle and management of white rust of mustard.
2. Describe the most distinguishing symptoms, causal organism, disease cycle and management of downy mildew disease of mustard.

(b) Briefly describe the most distinguishing symptoms and disease management of any of the following plant disease :

(i) Alternaria blight of mustard

(ii) Sclerotinia stem rot diseases of mustard

3. Describe white rust of mustard under the following headings.

(i) Pathogen (ii) Symptoms

(ii) Disease cycle (iv) Disease management

4. Describe in detail the symptoms, causal organism, disease cycle and management of powdery mildew of mustard.

5. Describe the symptoms, causal organism, disease cycle and the management of any disease of groundnut studied by you.

**Section 'B'**

**Short answer questions**

1. Give the chemical management of white blisters of white rust of mustard.
2. Distinguish between powdery and downy mildew of mustard.
3. Describe the hypertrophied symptoms of white blisters of mustard.

**Section 'C'**

**Very short answer questions**

1.Name the pathogen causing the following plant diseases:

(i) Downy mildew of mustard (ii) Powdery mildew of mustard

(iii) White blisters of mustard (iv) Alternaria blight of mustard

2.Name the disease caused by the following pathogens:

(i) *Albugo candida* (ii) *Peronospora parasitica*

(iii) *Alternaria brassicae* (iv) *Sclerotinia sclerotiorum*

3. Suggest cultural management of Sclerotinia stem rot disease of mustard.

4. Write down the etiology of *Peronospora parasitica* causing downy mildew of mustard.

5.Write down the chemical management of Alternaria blight of mustard.

**Section 'D'**

**Objective type questions**

Choose the correct answer from the following :

(i) Which one of the following diseases is caused by *Peronospora parasitica?*

a. Early leaf spot of groundnut b. Late leaf spot of groundnut
c. Wilt of linseed d. Downy mildew of mustard

(ii) Which one of the following pathogens causes white rust of mustard?
a. *Albugo candica* b. *Cercospora arachidicola*
c. *Sclerotinia sclerotiorum* d. *Melampsora lini*

(iii) Which Pathogen reproduce by sexual spores called oospores?
a. *Peronospora parasitica* b. *Mycospharella berkeleyii*
c. *Gibberella indica* d. None of the above

(iv) Which one of the following is an obligate parasite?
a. *Albugo candida* b. *Melampsora lini*
c. *Fusarium oxysporum* f.sp. *lini* d. *Mycosphaerella arachidis*

(v) Which one of the following has light olive coloured conidia with transverse and longitudinal septa?
(i) *Albugo candida* (ii) *Peronospora parasitica*
(iii) *Alternaria brassicae* (iv) *Sclerotinia sclerotiorum*

**Fig. 1:** White rust

**Fig. 2:** Downey mildew

**Fig. 2:** Alternaria leaf spot

**Fig. 4:** Sclerotinia Stem rot

**Plate 4:** Photograph showing symptoms of major diseases of mustard *(See colour version on page 204)*

# 5

# Diseases of Gram and Their Management

| Sl.No | Name of Diseases | Causal Organism |
|---|---|---|
| 1 | Wilt | *Fusarium oxysporum*f.sp. *ciceris* (Padwick) Matuoet. K Sato |
| 2 | Ascochyta blight | *Ascochyta rabiei* Pass Labr |
| 3 | Botrytis gray mold | *Botrytis cinerea* Pers.Ex.Fr |
| 4 | Rust | *Uromyces ciceris-arietini* Grogn. Jacz. & Beyer |
| 5 | Collar rot | *Sclerotium rolfsii* sacc |
| 6 | Dry root rot | *Rhizoctonia bataticola* Taub Butler |
| 7 | Wet root rot | *Rhizoctonia solani* |
| 8 | Alternaria blight | *Alternaria alternata* (Fr.) Keissl. and *Alternaria tenuis* |
| 9 | Anthracnose | *Colletotrichum dematium* |

## Wilt

### *Fusarium oxysporum* f.sp. *ciceris*

### Diagnostic Symptoms

- At first pale-green chlorosis, typically appear on lower leaves and extending up the plant.Leaves eventually take on a dull-yellow colour, wilt and the plant collapses and dies.
- Adult plants wilt, with their petioles and rachises drooping.In some cases there may be leaf vein clearing before wilt begins.
- Internally, the xylem tissues stain dark brown to almost black.
- Wilting may initially affect only one side of the plant( Fig.2).

### Etiology

### Scientific Classification

Kingdom : Fungi

Division : Eumycota

Sub- division : Deuteromycotina

Class : Hyphomycetes

Order : Hyphomycetales (Moniliales)

Family : Dematiaceae

Genus : *Fusarium*

Species : *oxysporum*f sp *ciceri*

- Mycelium: produces profusely branched septate hyphae, hyaline to light brown in color.
- Macroconidia :1-7 (mostly 3-5) septate, 25-55 to 2.5-6 µm, fusiform, curved, with a tapering, pointed, sometimes hooked, apical cell and distinctly pedicellate basal cell. In some strains macroconidia are sparse, in others, abundantly produced in pinkish sporodochial conidiomata.
- Microconidia: borne on simple short conidiophores arising laterally on the hyphae.These are oval to cylindrical, straight to curved, and 5-15 to 2-5 µm, collecting in small, slimy droplets.
- Chlamydospores: globose, hyaline 7-15 µm diameter, usually forming abundantly in mature colonies, terminal or intercalary, single or in small groups or chains.

**Disease Cycle**

- Pathogen is mainly soil borne as well as seed borne and also it is a facultative saprophyte.It can survive in the soil upto six years in the absence of susceptible host.
- Primary infection is through chlamydospores in soil, which remain viable upto next crop season.
- Secondary spread is through irrigation water, cultural operations and implements.
- Following infection of host roots, the fungus crosses the cortex and enters the xylem tissues.
- It then spreads rapidly up through the vascular system, becoming systemic in the host tissues, and may directly infect the seed

**Favorable Conditions**

- High soil temperature – 25°C.
- High soil moisture.

**Integrated Management**

- Grow resistant varieties like Avrodhi, Alok Samrat, Pusa-212, JG- 322 GPF-2, Haryanachana-1 and Kabuli chickpea like Pusa-1073 and Pusa-2024, ICCC 42, H82-2.

- Seed treatment with Carbendazim 1 g+Thiram 1g/kg seeds. Treatment with *Trichoderma viride* at 4 g/kg or *Pseudonomas fluorescens* @ 10g/kg of seed.
- Apply heavy doses of organic manure or green manure.

## Ascochyta Blight

### *Ascochyta rabiei*

### Diagnostic Symptoms

- All above ground parts of the plant are infected.
- Round or elongated lesions, bearing irregularly depressed brown spot and surrounded by a brownish red margin on leaf.
- Similar spots may appear on the stem and pods.
- Spots on the stem and pods have pycnidia arranged in concentric circles as minute block dots.
- When the lesions girdle the stem, the portion above the point of attack rapidly dies.
- If the main stem is girdles at the collar region, the whole plant dies (Fig.1).

### Etiology

### Scientific Classification

Kingdom : Fungi

Division : Eumycota

Sub- division : Deuteromycotina

Class : Coelomycetes

Order : Sphaeropsidales

Family : Sphaeropsidaceae

Genus : *Aschochya*

Species : *rabiei*

- Pathogen occurs as both an anamorph (nonsexual state) and teleomorph (sexual state).
- Anamorph: *Ascochyta rabiei*
- Mycelium: hyaline to brown, septate.

- Pycnidia: fruiting body formed on stems, leaves and pods, round or flattened, globose, dark brown, immersed in the tissues, measuring 100 to 200 µm in diameter, with a prominent ostiole are spherical to sub-globose.
- Conidia: formed inside pycnididum,hyaline, oval to oblong, straight or slightly curved, 1-septate, some 0-septate, slightly or not constricted at the septum, rounded at each end and measure 10-16x3.5 µm in size.
- Teleomorph:*Didymella rabiei(Mycosphaerella rabiei)*
- In perfect stage, the dark colored, globose pseudothecia are formed , each measuring 100 to 140 µm.
- Pseudothecia are characterized by eight ascospore asci.The ascospores are bicelled with one cell larger than the other,measure 12.5-19.0x6.7 -7.6 µm.

## Disease Cycle

- Pathogen survives in the infected plant debris as pycnidia or in seed.
- Primary infection is from inoculums produced by pathogen from seed-borne pycnidia and plant debris in the soil.
- Secondary infection is mainly through air-borne conidia.
- Rain splash also helps in the spread of the disease.

## Favorable Conditions

- Temperature 20-25°C.
- Relative humidity -60%.
- High rainfall during flowering.
- Leaf wetness -17h.

## Integrated Management

- Use only certified seed.
- Use resistant varieties *viz.* Him chana -1, Gaurav, Vardan, Samrat, PBG 1 and BG 261
- Treat the seeds with Thiram + Carbendazim (1:1 ratio) at 2 g/kg.
- Exposure of seed at 40-50°C reduced the survival of *A. rabiei* by about 40-70 per cent.
- Foliar spray with Chlorothalonil (0.15%) solution, if required repeat after 15 days.
- Destroy blight-infested crop residues and volunteer chickpea plants.

- Choose crop rotations that permit 3 to 4 years between successive chickpea crops.

## Botrytis Gray Mold

### *Boirytis cinerea* pers. Ex. Fr

### Diagnostic Symptoms

- Disease is usually seen at flowering time when the crop canopy is fully developed.
- Lack of pod setting is the first indication of the disease.
- Under favorable conditions foliage shows clear symptoms and plants often die in patches.
- More severe on portions of the plant hidden under the canopy.
- Shed flowers and leaves, covered with the spore mass, can be seen on the ground under the plants.
- When humidity is very high, the symptoms appear on stems, leaves, flowers and pods as gray or dark brown lesions covered with moldy sporophores.
- Lesions on stem girdle the stem completely .
- Tender branches break off at the point where grey mold has caused rotting.
- Affected leaves and flowers turn into a rotting mass.
- Lesions on the pod are water soaked and irregular.
- Grayish white mycelium may be seen on the infected seeds (Fig.3).

### Etiology

- Mycelium: septate, brown, hyphae being 8-16µ m wide.
- Conidiophores: light brown, septate, erect and their tips are slightly enlarged bearing small and pointed sterigmata.
- Conidia: hyaline, 1-celled, oval and are borne in clusters on short sterigmata.Conidia in mass are ash-grey in colour, and measure 4-16 x 4-20 µm.

### Disease Cycle

- Pathogen survives in the infected plant debris. However, infected seed is thought to be the primary source of infection.
- Pathogen is a facultative parasite with a very wide host range in the temperate and sub-tropical region.

- It causes gray mould disease in a number of crop plants, such as strawberry, grapevine, apple, cabbage, carrot, cucumber, eggplant, lettuce, pepper, squash, tomato and several ornamentals.
- Pathogen also produces highly resistant sclerotia as survival structures in older cultures. It overwinters as sclerotia or intact mycelia, both of which germinate in spring to produce conidiophores.
- Spores of *B. cinerea* are released from conidiophores and then dispersed by air currents or rain splash and cause new infections.

### Favorable Conditions

- Optimum temperature- 28 to 30 $^{0}$C
- Relative humidity – 95-100 per cent

### Integrated Management

- Grow resistant varieties: BG 276, GL 90159, GL 91040, GL 91071, and GL 92162
- Adopt wider spacing.
- Intercropping with linseed.
- Avoid excessive vegetative growth.
- Avoid excessive irrigation.
- Growing compact varieties.
- Seed treatment with Carbendazim + Thiram @ 3 g/kg.
- Foliar spray of @ g/L, if required repeat after 15 days.

## MODEL QUESTION PAPER

### Section 'A'

### Long answer questions

1. Describe in detail the symptoms, etiology, disease cycle and management of Ascochyta blight of chickpea.
2. Briefly describe the most distinguishing symptoms and disease management of any two of the following plant disease :

   (i) Collar rot of chickpea (ii) Dry root rot of chickpea

   (ii) Wilt of chickpea
3. Describe the wilt of chickpea in the following headings :

   (i) Pathogen (ii) Symptoms

   (ii) Disease cycle (iv) Disease management

4. How would you identify the following diseases in field :
   (i) Collar rot of chickpea (ii) Dry root rot of chickpea
   (ii) Wilt of chickpea (iv) Wet root rot of chickpea
5. Give the disease cycle and management of the following diseases :
   (i) Wilt of chickpea (ii) Botrytis grey mould of chickpea
6. Describe in detail the most distinguishing symptoms, causal organism, disease cycle and management of botrytis grey mould of chickpea.

**Section 'B'**

**Short answer questions**

1. Compare and contrast wet root rot and dry root rot of chickpea giving suitable measures for their management.
2. Give the disease cycle and predisposition of Ascochyta blight of chickpea.
3. Give the Integrated management schedule of Botrytis grey mould of chickpea.

**Section 'C'**

**Very short answer questions**

(i) Suggest some disease resistant varieties of chickpea against with disease.
(ii) Name the diseases caused by the following pathogens:
   (a) *Fusarium oxysporum f.* sp. *ciceri* (b) *Sclerotium rolfsii*
   (c) *Uromyces ciceris-arietini* *(d) Rhizoctonia solani*
(iii) List some important diseases and their causal organisms of gram.
(iv) Distinguish between wilt and collar rot of gram.
(v) Name the fungal pathogens causing the following diseases:
   (a) Collar rot of gram (b) Rust of chickpea
   (c) Wet root rot of gram (d) Aschochyta blight of gram

**Section 'D'**

**Objective type questions**

(a) Fill in the blanks with suitable word (s)
   (i) *Fusarium oxysporum f.* sp. *ciceri* ......................... of chickpea.
   (ii) Wet root rot of gram is caused by ........................
   (iii) .................. is the perfect state or teleomorph of *Ascochyta rabiei.*

(iv) Chlamydospore is the resting spore of the fungus ...............

(v) Pycnidia is fruiting body found in fungus ...................

(b) State whether the following statements are True or False.

(i) The pathogen causing wilt in chickpea is mainly soil borne.

(ii) *Gibberella indica* is the perfect state or teleomorph of *Ascochyta rabiae.*

(iv) Lack of pod setting is the first indication of the BGM disease.

(iv) The fungus *Fusarium oxysporum* f sp. *ciceri* produces three types of conidia like macroconidia, microconidia and chlamydospores

(v) Wilt disease of crops can easily be controlled by protective sprays of fungicides.

**Fig. 1:** Ascochyta blight

**Fig. 2:** Wilt

**Fig. 3:** Botrytis grey mold

**Plate 5:** Photograph showing symptoms of major diseases of chickpea *(See colour version on page 204)*

# 6

# Diseases of Lentil and Their Management

| Sl.No | Name of Diseases | Causal Organism |
|---|---|---|
| 1 | Wilt | *Fusarium oxysporum* (Schlecht) Snyder & Hansen |
| 2 | Rust | *Uromyces fabae* (Pers.) Schröt. |
| 3 | Anthracnose | *Colletotrichum truncatum* (Schwein.) Andrus & Moore |
| 4 | Ascochyta blight | *Ascochyta lentis* Jellis & Punith |
| 5 | Botrytis gray mold | *Boitrytis cinerea* |
| 5 | Sclerotinia rot/collar rot | *Sclerotinia rolfsii Sacc.* |
| 7 | Powdery mildew | *Erysiphe pisi* DC. |

## Wilt

### *Fusarium oxysporum f.*sp. *lentis*

### Diagnostic Symptoms

- Symptoms can occur at both the seedling and plant developmental stages and appear as patches in the field
- It usually occurs near or at the reproductive stages (flowering to pod-filling) of crop growth.
- Symptoms include the drooping and wilting of the uppermost leaflets .
- Plants become completely yellow and die.
- Plants are affected during the mid- to late-pod filling stages, seeds are often noculums
- Root system appears healthy, but with a reduced proliferation and nodulation rate.
- Longitudinal section of the stems show a brownish red discoloration of the internal tissues, first at the base ,later up the stem (Fig.1).

### Etiology

### Scientific Classification

Kingdom : Fungi

Division : Eumycota

Sub- division : Deuteromycotina

Class : Hyphomycetes

Order : Hyphomycetales (Moniliales)

Family : Dematiaceae

Genus : *Fusarium*

Species : *oxysporum*

- Mycelium: hyaline ,spetate, much branched and range in colour from white to pale and violet.
- Microconidia: abundant, aseptate, reni form to oval, ranging from 5-12 μm ×2.-3.5 μm produced in false heads on short monophialide conidiophores.
- Macroconidia:Usually two to seven celled, long with pointed apical cell and notched basal cell.
- Chlamydospore: single celled, oval or spherical shaped and thick walled, formed singly in macroconidia or apical or intercalary in the hyphae

**Disease Cycle**

- Pathogen is soil borne and survive in a dormant stage (as chlaymodsopres) in the soil for several years.
- The pathogen can also survive within infected plant material in the field.
- This indicates that the pathogen is well adapted to survive adverse conditions.

**Favorable Conditions**

- Soil temperature -23° to 27°C.
- Moist soil and warm and humid conditions

**Integrated Management**

- Use resistant varieties (PL 406, PL 639 and PL 234) .
- Seed treatment with Carbendazim + Thiram at (1+1) 2 g/kg or Seeds treatment with *Trichoderma viride* at 4 g/kg or *Pseudonomas fluorescens* @ 10g/kg of seed
- Follow crop rotation.
- Deep ploughing over summer and removal of infected trash can reduce inoculums levels.

- Solarization of soil by covering the soil with transparent polythene sheet (75 gauge) for 6-8 weeks during the summer.
- Apply heavy doses of organic manure or green manure.
- Remove all weeds from the field.
- Crop rotation of up to 5-7 years to reduce levels of inoculums in the soil.

## Rust

### Diagnostic Symptom

- Intially small white slightly raised spots appear on the upper surface of the leaves.
- As they enlarge change to orange brown in color, often surrounded by light colored halo.
- These pustules are to be found on both upper and lower side of leaves, stems and pods.
- In severe infection, the plant dries up without forming seed in pods or shriveled seeds are formed (Fig. 2).

### Etiology

### Scientific Classification

| | |
|---|---|
| Phylum | : Basidiomycotina |
| Class | : Teliomyetes |
| Order | : Uredinales |
| Family | : Pucciniaceae |
| Genus | : *Puccinia* |
| Species | : *Striiformis* |

- Automacrocyclic : all five spore states (spermagonial, aecial, uredinial, telial and basidial) occur, and there are no alternate hosts.
- Spermagonia: mostly on abaxial leaf surface, amphigenous in small groups associated with aecia.
- Aecia: mostly on abaxial leaf surface in small groups, predominantly along veins surrounding the spermagonia or sometimes scattered, peridium cupulate, whitish, 0.3–0.4 mm diam.; aeciospores 18–26 × 15–21 μm, broadly ellipsoid, wall hyaline (colorless), finely verrucose and 1–1.5 μm thick.
- Uredinia:amphigenous, yellowish brown (cinnamon), 0.5 mm diam.; urediniospores 22–32 × 17–25 μm, broadly ellipsoid, wall light golden brown, 1–2.5 μm thick, uniformly echinulate, pores three to five, equatorial or occasionally scattered.

- Telia: sometimes on adaxial surface or sometimes amphigenous and on stems, exposed, blackish brown, compact and 1–2 mm diam; teliospores ellipsoidal, obvoidal or noculums, rounder or subacute above, 24–40 × 17–26 μm; wall chestnut-brown, smooth, 1–3 μm thick at the sides, 5–12 μm thick at the apex, pedicels brownish at least apically, up to 100 μm long.

## Disease Cycle

- Pathogen survives on plant debris, volunteer plants and weeds when crops are available. It can also carried on seeds as concomitant contaminations.
- Commonly, the telial state has a survival value. Teliospores produced by a telium to survive unfavorable periods.
- These teliospores can germinate and produce basidia and basidiospores capable of infecting lentils and starting the infection cycle.
- Secondary spread by wind over large distance to infect new plants or fields.

## Favourable conditions

- Temperature- 17- 25°C
- Prolonged leaf wetness.

## Integrated Management

- Use resistant varieties (PL 234, PL 406, LL 931, LL 699 and Narendra Masur-1).
- Summer deep ploughing.
- Follow crop rotation with non hostcrops.
- Field sanitation, rogueing.
- Destroy the alternate host plants.
- Foliar spray Copper oxychloride,(0.3%) Difenoconazole (0.1%) or Chlorothalonil (0.2%), if required repeat after 15 days.

## MODEL QUESTION PAPER

### Section 'A'

### Long answer questions

1. Explain in detail the symptoms, etiology, disease cycle and management of wilt of lentil.

2. Briefly describe the most distinguishing symptoms and disease management of any two of the following lentil disease :
   (i) Collar rot (ii) Rust
   (iii) Wilt
3. Describe the Anthracnose of lentil in the following headings:
   (i) Pathogen (ii) Diagnostic symptoms
   (ii) Disease cycle (iv) Disease management
4. How would you identify the following diseases of lentil in field:
   (i) Collar rot (ii) Rust
   (ii) Wilt (iv) Anthracnose
5. Give the disease cycle and management of the following lentil diseases:
   (i) Ascochyta blight (ii) Botrytis grey mold
6. Describe in detail the most distinguishing symptoms, causal organism, disease cycle and management of rust of lentil.

**Section 'B'**

**Short answer questions**

1. Compare and contrast collar rot and wilt of lentil giving suitable measures for their management.
2. Give the disease cycle and predisposition of ascochyta blight of lentil.
3. Give an Integrated management schedule of wilt of lentil.

**Section 'C'**

**Very short answer questions**

(i) Suggest some disease resistant varieties for lentil wilt disease.
(ii) List some important diseases and their causal organisms of lentil.
(iii) Distinguish between symptom of wilt and collar rot of lentil.
(iv) Name the fungal pathogens causing the following diseases of lentil :
   (a) Collar rot (b) Rust
   (c) Root rot (d) Wilt

## Section 'D'

## Objective Type Questions

(a) Fill in the blanks with suitable word (s)

(i) *Fusarium oxysporum f.* sp. *lentis* is causal organism of .............. of lentil.

(ii) Rust of lentil is caused by .........................

(iii) .................. is the perfect state or teleomorph of *Ascochyta rabiei.*

(iv) Chlamydospore is the resting spore of the fungus ..............

(v) Pycnidia is fruiting body found in fungus ...................

(b) State whether the following statements are *True* or *False.*

(i) The pathogen causing wilt in lentil is mainly soil borne.

(ii) *Gibberella indica* is the perfect state or teleomorph of *Ascochyta rabi.*

(iii) Lack of pod setting is the first indication of the BGM disease.

(iv) The fungus *Fusarium lentis* produces three types of conidia like macroconidia, microconidia and chlamydospores

(v) Wilt disease of crops can easily be controlled by protective sprays of fungicides.

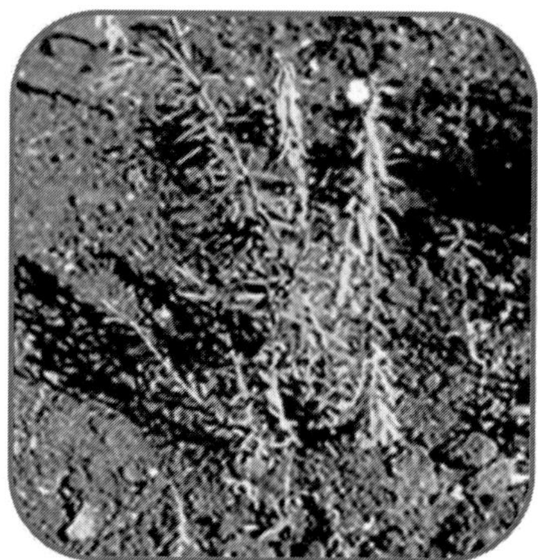

**Fig. 1:** Wilt

**Fig. 2:** Rust

**Plate 6:** Photograph showing symptoms of major diseases of lentil *(See colour version on page 205)*

# 7

# Diseases of Cotton and Their Management

| Sl.No | Name of Diseases | Causal Organism |
|---|---|---|
| 1 | Wilt | *Fusarium oxysporum* f.sp. *vasinfectum* (Akt) Synedar & Hansen |
| 2 | Anthracnose | *Colletotrichum capsici* |
| 3 | Black arm/Angular leaf spot/Bacterial blight | *Xanthomonas axonopodis* pv. *malvacearum*(Smith) Dye |
| 4 | Leaf blight | *Alternaria macrospora* Zimm |
| 5 | Leaf Curl | *Cotton leaf curl virus* |

## Wilt

### *Fusarium oxysporum* f.sp. *vasinfectum*

### Diagnostic Symptoms

- Disease affects the crop at all stages.

### Seedling stage

- Cotyledons of the seedling turn yellow and then brown.
- Base of petiole shows brown ring, followed by wilting and drying of the seedlings.

### Young and grown up plants

- Symptoms start from the older leaves at the base, followed by younger ones towards the top, finally involving the branches and the whole plant.
- Initial symptom is yellowing of edges of leaves and area around the veins i.e. discoloration starts from the margin and spreads towards the midrib.
- The leaves loose their turgidity, gradually turn brown, droop and finally drop off.
- Defoliation or wilting may be complete leaving the stem alone standing in the field.

- Sometimes partial wilting occurs; where in only one portion of the plant is affected, the other remaining free.
- Browning or blackening of vascular tissues is the other important symptom,
- Black streaks or stripes may be seen extending upwards to the branches and downwards to lateral roots.
- In severe cases, discolouration may extend throughout the plant starting from roots extending to stem, leaves and even bolls.
- In transverse section, discoloured ring is seen in the woody tissues of stem.
- Plants affected later in the season are stunted with fewer bolls which are very small and open before they mature (Fig-1).

## Etiology

### Scientific Classification

Kingdom : Fungi

Division : Eumycota

Sub- division : Deuteromycotina

Class : Hyphomycetes

Order : Tuberculariales

Family : Tuberculariaceae

Genus : *Fusarium*

Species : *oxysporum* f. sp.*vasinfectum*

- Mycelium: white to grayish white or bluish purple, often form a mat on the collar region of the stem near the ground level.
- Hyphae: both inter as well as intracellular .
- Conidiophores: produced in sporodochia ,sometimes on the mycelium, vertically branched.
- Conidia: macrocondia, microconidia and chlamydospores.
- Macroconidia: 1 to 5 septate, hyaline, thin walled, falcate, 40-50x3-4.5μm.
- Microconidia: hyaline, thin walled, spherical or elliptical, unicellular, measuring 5-12x 2-3.5 μm.
- Chlamydospores: dark coloured and thick walled, terminal or intercalary

## Disease Cycle

- It is a soil inhabitant and has good saprophytic ability.
- Pathogen can survive in soil as saprophyte and as chlamydospores for many years.
- Pathogen is both externally and internally seed-borne.
- Primary infection is mainly through conidia produced from dormant hyphae and chlamydospores in the soil.
- Secondary infection in case of this disease is rare as the conidia produced seldom succeed in establishing aerial infection on the plant.

## Favorable Conditions

- Soil temperature – 20-30°C.
- Soil humidity- 40 to 70%.
- pH=5.3.
- Hot and dry periods followed by rains.
- Heavy black soils with an alkaline reaction.
- Increased doses of nitrogenous fertilizers.

## Integrated Management

- Deep summer ploughing during June-July
- Grow disease resistant varieties like Varalakshmi, Vijay Pratap, Jayadhar and Verum.
- Soak seed in 40-50 ppm streptocyclin solution before sowing.
- Treat the acid delinted seeds with Carboxin + Thiram at 2 g/kg.
- Foliar spray with a mixture of Streptocyclin (Streptycyclin sulphate 90%+Tetracyclin hydrochloride 10%SP) on the appearance of field symptoms, if required repeat 10-15 days interval.
- Remove and burn the infected plant debris in the soil.
- Apply heavy doses of farm yard manure or other organic manures.
- Follow mixed cropping with non-host plants.
- Spot drench with Carbendazim + Mancozeb @ 2 g/litre.
- Delay planting to the end of October.
- Avoid dense planting.
- Retain cotton residues on the surface for 60 days.
- Bare fallow rotation is best.
- Always practice good farm hygiene.

## Anthracnose

*Colletotrichum capsici*

### Diagnostic Symptoms

- Pathogen infects the seedlings and produces small reddish circular spots on the cotyledons and primary leaves.
- Lesions develop on the collar region, stem may be girdled, causing seedling to wilt and die.
- In mature plants, the fungus attacks the stem, leading to stem splitting and shredding of bark.
- Most common symptom is boll spotting. Small water soaked, circular, reddish brown depressed spots appear on the bolls.The lint is stained to yellow or brown, becomes a solid brittle mass of fiber.The infected bolls cease to grow and burst and dry up prematurely (Fig-2).

### Etiology

#### Scientific Classification

Kingdom : Fungi

Division : Eumycota

Sub- division : Deuteromycotina

Class : Coelomycetes

Order : Melanconiales

Family : Melanconiaceae

Genus : *Colletotrichum*

Species : *capsici*

- Pathogen forms large number of acervuli on the infected parts.
- Conidiophores: are slightly curved, short and club shaped.
- Conidia are hyaline and falcate, borne single on the conidiophores.
- Setae: Numerous black colored and thick walled setae are also produced in acervulus.

### Disease Cycle

- Pathogen survives as dormant mycelium in the seed or as conidia on the surface of seeds for about a year.

- Pathogen also perpetuates on the rotten bolls and other plant debris in the soil.
- Secondary spread is by air-borne conidia.

**Favorable Conditions**

- Prolonged rainfall at the time of boll formation.
- Close planting.

**Integrated Management**

- Remove and burn the infected plant debris and bolls in the soil.
- Rogue out the weed hosts.
- Treat the delinted seeds with Carboxin + Thiram @ 3.5/kg seed.
- Foliar spray with Carbendazim+Mancozeb (0.25%) or Copper oxychloride (0.3%) at boll formation stage of crop, if required repeat after 15 days.

**Black Arm/Angular leaf Spot**

***Xanthomonas axonopodis* pv. *malvacearum***

**Diagnostic Symptoms**

- Pathogen attacks all stages from seed to harvest.
- Small water soaked spots, angular lesins of 1 to 5 mm across the leaves and bracts, especially on the undersurface of leaves. Hence called angular leaf spot.
- Sometimes extensive dark green, water soaked lesions along the veins known as veinblight.
- Lesions dry and darken with age and leaves may be shed prematurely resulting in extensive defoliation.
- Black lesions on the stem which girdle and spread along the stem or branch known as black arm.
- Dark green, water soaked, greasy, circular lesions of 2 to 10mm across the bolls, especially at the base of the boll under the calyx crown.
- As the boll matures the lesions dry out and prevent normal boll opening. This phase of symptom is called as "Boll rot"
- The bacterium spreads inside the boll and lint gets stained yellow because of bacterial ooze and loses its appearance and market value.
- The pathogen also infects the seed and causes reduction in size and viability of the seeds(Fig-3).

## Etiology

### Scientific Classification

Kingdom : Prokaryotae

Division : Gracilicutes

Class : Proteobacteria

Family : Psudomonadaceae

Genus : *Xanthomonas*

Species : *axonopodis* pv. *malvacearum*

- Gram negative.
- Short rod with a single polar flagellum.
- Non-spore forming measures 1.0-1.2 X 0.7-0.9 μm.
- In culture, it forms yellow colony

### Disease Cycle

- Pathogen inoculums may either be present in the field on infected crop residues from a previous season or it may be introduced at planting within infected seed.
- Infected seeds are the source of primary inoculums.They carry the pathogen externally as well as internally.
- Plant debris is another source of inoculums
- Lesions on cotyledons may be initiated by inoculums within the seed during germination.
- Inoculum from infected crop residues may be splashed onto the foliage and into the growing point of young seedlings where it can survive saprophytic ally on leaf surfaces.
- When environmental conditions are favorable the bacteria enter the plant via the stomata or wounds.
- As lesions develop bacteria exude out onto the leaf surface for further dispersal through wind driven rain.
- Pathogen is able to enter the seed when mature, open, blight-infected bolls are exposed to wet weather prior to harvest.

### Favorable Conditions

- High atmospheric temperature – 30-40°C.
- Relative humidity – 85 per cent.

- Optimum soil temperature – 28°C.
- Early sowing.
- Delayed thinning.
- Poor tillage and late irrigation.

## Integrated Management

- Grow resistant varieties like CRH 71 and Sujatha, 1412
- Delint the cotton seeds with concentrated sulphuric acid at 100ml/kg of seed.
- Soak the seeds in Streptomycin sulphate (0.1%) overnight or treat the delinted seeds with Carboxin or Oxycarboxin at 2 g/kg.
- Foliar spraywith Streptomycin sulphate_+Tetracycline mixture 100g along with Copper oxychloride at 1.25 Kg/ha.
- Remove and destroy the infected plant debris.
- Rogue out the volunteer cotton plants and weed hosts.
- Follow crop rotation with non-host crops.
- Early earthing up with potash.

## MODEL QUESTION PAPER

## Section 'A'

## Long answer questions

1. Describe in detail the occurrence importance, symptoms, etiology, disease cycle and management of wilt of cotton.
2. Give the diagrammatic representation of disease cycle of the following diseases of cotton

   (i) Wilt (ii) Anthracnose

   (iii) Bacterial blight (iv) Leaf curl
3. Briefly describe the diagnostic symptoms and integrated management of any two of the following cotton diseases:

   (i) Collar rot (ii) Bacterial blight

   (ii) Wilt
4. Illustrate the Angular leaf spot of cotton in the following headings :

   (i) Pathogen (ii) Symptoms

   (ii) Disease cycle (iv) Disease management

5. Describe in detail the most distinguishing symptoms, causal organism, disease cycle and management of para wilt of cotton.

## Section 'B'

### Short answer questions

1. Compare and contrast bacterial blight and leaf blight of cotton giving suitable measures for their management.
2. Give the disease cycle and predisposition of bacterial blight of cotton.
3. Write an Integrated management scheduele of angular leaf spot of cotton.

## Section 'C'

### Very short answer questions

i. Suggest some disease resistant varieties of cotton against will disease.

(ii) Mention the name the resting structures by the following pathogens:
   - (a) *Fusarium oxysporum f.* sp. *vasinfectum*
   - (b) *Alternaria macrospora*
   - (c) *Colletotrichum capsici*
   - *(d) Rhizoctonia solani*

(iii) List some important diseases and their causal organisms of cotton?

(v) Distinguish between fungal wilt and bacterial blight of cotton?

(vi) Which type of disease is can be effectively managed by crop rotation ?

## Section 'D'

### Objective Type questions

(a) Fill in the blanks with suitable word (s)

   (i) *Xanthomonas axonopodis* pv. *malvacearum* is a cotton ... .................. bacteria

   (ii) Anthracnose of cotton is caused by ........................

   (iii) Numerous black colored and thick walled ..................... are also produced in acervulus of Colletotrichum spp.

   (vii) Chlamydospore is the resting spore of the fungus ..............

   (viii) The fungus *Fusarium oxysporum* f.sp. *vasinfectum* produces three types of conidia like ................, .............. and ...................

(b) State whether the following statements are True or False.

(i) *Xanthomonas axonopodis* pv. *malvacearum* forms yellow colony in culture.

(ii) Foliar spray with Streptomycin sulphate_+Tetracycline mixture 100g along with Copper oxychloride at 1.25 Kg/ha.

(iii) Black lesions on the stem of cotton which girdle and spread along the stem or branch known as black arm.

(iv) The fungus *Fusarium oxysporum* f.sp. *vasinfectum* produces three types of conidia like macroconidia, microconidia and chlamydospores

(v) Wilt disease of cotton can be managed by soil application of Trichoderma.

**Fig. 1:** Fusarium wilt

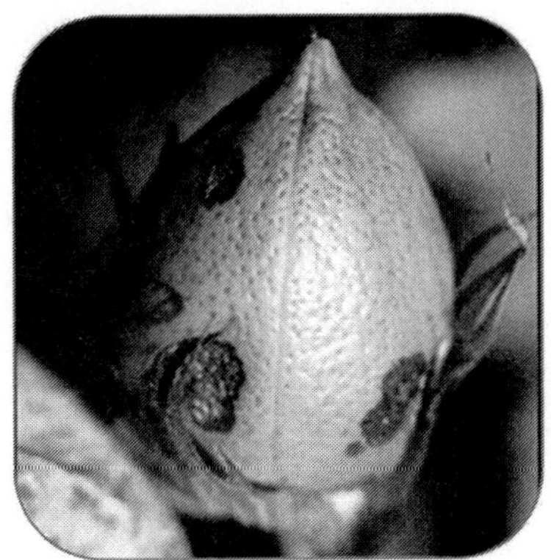

**Fig. 2:** Anthracnose

**Fig. 3:** Bacterial blight

**Plate 7:** Photograph showing symptoms of major diseases of cotton *(See colour version on page 206)*

# 8

# Diseases of Pea and Their Management

| Sl.No | Name of Disease | Causal Organism |
|---|---|---|
| 1 | Powdery mildew | *Erysiphe pisi* DC. Syn. *E. polygoni* DC. |
| 2 | Downy mildew | *Peronospora pisi* Sydow |
| 3 | Rust | *Uromyces fabae (Pers.)* Schrot. & *Uromyces fabae DB.* f. sp. *pisis ativae* Hiratsuka |
| 3 | Anthracnose | *Colletotrichum pisi* |
| 4 | Fusarium wilt | *Fusarium oxysporum* f. sp. *pisi* |

## Powdery Mildew

### *Erysiphe polygoni*

### Diagnostic Symptoms

- At first symptom appears on leaves.
- White floury patches appear on both surface of leaves as well as tendrils, stem and pod.
- As plant become older, the symptoms almost cover the entire plant, become more or less grayish brown and the infected parts impart dirty appearance.
- Fruits do not either set or remain very small.
- Later stages, powdery growth also covers the pods(Fig-1).

### Etiology

### Scientific Classification

Kingdom : Fungi

Division : Eucomycota

Division : Ascomycotina

Class : Pyrenomycetes

Order : Erysiphales

Family : Erysiphaceae

Genus : *Erysiphe*

Species : *polygoni*

- Pathogen: obligate parasite and ectophite, only haustoria enter into the cell.
- Mycelium: septate, hyaline, profusely branched, and superficial forming a white web like coating over the surface of the infected part.It sends finger like haustoria into the epidermal cells of the host to obtain nutrients.After maturity, many hyphal branches give rise to erect, stout, conidiophores, which produce conidia in chain.
- Conidiophores:bears spores in chain , septate and their cells appear like conidia in shape.
- Conidia: elliptical or barrel shaped, hyaline, single celled, and measures 25-35x13-16 μm.
- Cleistothecia: black, minute bodies provided with a number of appendages, and remain scattered in the mycelia web, contains usually 2-8 asci.
- Ascospores: 3-8 , hyaline, one celled, elliptical and unicellular, measuring 19-25x9-14μm.

## Disease Cycle

- Pathogen survives in soil and plant debris in the form of cleistothecia and may survive in seed in the form of dormant mycelium.
- Primary infection occurs by asci and ascospores releases after breakdown of cleistothecia or conidia from collateral hosts.
- Secondary spread takes place by wind borne conidia produced as a result of primary infection.

## Favorable Conditions

- Optimum temperature-15-25°C.
- Relative humidity -over 70%.
- Absense of rain.

## Integrated Management

- Field sanitation.
- Growing of early maturing and resistant varieties like Rachna, Pant P5, DMR 11, HUP 2, JP 885, KFP 103, Ambika, Shubhra, Aparna, Azad P4, Pusa Panna

- Foliar spray with Azoxystrobin +Difenoconazole SC @1 ml/l or Wettable Sulphur 40% WP 0.2% or Triadimefon 25% WP @ 0.1 % . Second spray after 25 days of interval.
- Control volunteer field peas.
- Avoid sowing field pea crops adjacent to last season's stubble.
- Burn infected pea stubble soon after harvest.

## Downy Mildew

***Peronospora pisi***

### Diagnostic Symptoms

- At first grayish white, moldy growth appears on the lower leaf surface, and a yellowish area appears on the opposite side of the leaf.
- Infected leaves can turn yellow and die if weather is cool and damp.
- Stems may be distorted and stunted.
- Brown blotches appear on pods, and mold may grow inside pods(Fig-2).

### Etiology

### Scientific Classification

Phylum : Heterokontophyta

Class : Oomycota

Order : Peronosporales

Family : Peronosporaceae

Genus : *Peronospora*

Species : *pisi*

- Pathogen is an obligate parasite.
- Mycelium: coenocytic, hyaline, profusely branched, intercellular sending branched, finger shaped haustoria inside the host cells.
- Conidiophores: unbranched for about two third or more of their length and dichotomously branch at the tip.Branches are long, slender and pointed and are called sterigmata.A single conidium is produced at the tips of each sterigmata.
- Conidia: oval to elliptical and light yellow in color.Conidia are short lived and germinate readily by germ tube.
- Antheridia & Oogonnia: When the pathogen moves towards the, end of the growing season, its hyphae move deep into the host tissue

inter,spaces wherein they produce sex organs (antheridia and oogonia) in close proximity.

- Oospore: After fertilization via fertilization tube, thick walled, green yellow, warty oospores are produced at the rate of one in each oogonium.

## Disease Cycle

- Pathogen survives in the soil and on old pea trash through oospores.It can also be seed-borne.
- Oospores perennating in soil germinate by germ tube at the return of favorable conditions.
- The germ tube infects underground parts of young seedlings and the pathogen progresses upwardly systemically causing primary infection.
- These diseased plants provide the wind blown spores that spread the disease in the field and serve as source of secondary infection.
- The oogonial and oospore stages occur later; before harvest, the oospores finally reaches the soil, where they survive one to two years.

## Favorable Conditions

- Temperature- 10-20 ºC.
- Relative humidity- 90%.

## Integrated Management

- Field sanitation.
- Growing of resistant varieties.
- Seed treatment with Metalaxyl at 2.5g/kg.
- Foliar spray with Metalaxyl + Mancozeb @ 0.2%, if required repeat after 15 days of first spraying.
- Deep ploughing of fields during summer.
- Timely sowing should be done.
- Destroy the alternate host plants.
- Apply manures and fertilizers as per soil test.
- Destruction of diseased plant debris.

## Rust

***Uromyces pisi* (Pers.) Schrot.**

## Diagnostic Symptoms

- Disease appears usually at the beginning of flowering.

- Leaves of infected plants exhibit many small, orange-brown pustules usually at the lower surface.
- Severely infected leaves wither and may drop from the plant.
- Larger pustules occur on the stems and isolated pustules may be found on the pods(Fig-3).

## Etiology

Phylum : Basidiomycotina
Class : Teliomyetes
Order : Uridinales
Family : Pucciniaceae
Genus : *Uromyces*
Species : *pisi*

- Autoecious rust:Pathogen passes all stages on peas completing its life cycle on pea.
- Aecial stage: appear first on the lower surface of leaves and on the stems and petioles, where the spermogonia also occur mixed with aecia.
- Aeciospores:are elliptical, yellowish brown, measure 14-22μ in diameter and have finely warty cell walls.
- Uredial stage: which is repeated several times during the season, produces spiny light brown uredospores.
- Urediniospores: are rounded, single, spinose, light brown, 21-30 x 18-26 microns.Pathogens produce several generations of urediniospores on peas.
- Telial stage: are not powdering, dark-brown, almost black.
- Teliospores: are unicellular, rounded, glabrous, positioned on colorless leg, 25-40 x 18-28 microns in size.It germinate to produce four celled basidia.
- Basidiospores: on which four sporidia are formed.
- Spores are spread by means of wind.

## Disease Cycle

- Pathogen survives in soil and plant debris.
- Primary infection on the newly growing crop is considered to be caused by the germination of teleutospores present in soil, seed and collateral hosts in the month of January.
- Secondary infection by aeciospores or secondary aecia.

## Favorable Conditions

- Frequent rainfall.
- Plentiful dews.
- Air temperature- 20-25°C

## Integrated Management

- Field sanitation.
- Growing of resistant varieties.
- Early sowing.
- Destruction of diseased plant debris.
- Crop rotation.
- Foliar spray of Carbendazim+ Mancozeb (0.25%) or Triadimefon 25% WP (0.1 %)given at 15 days intervals starting from first week of December.

## MODEL QUESTION PAPER

## Section 'A'

## Long answer questions

1. Describe the causal organisms and mode of penetration of powdery mildew and downy mildew diseases of pea.
2. Describe in detail the occurrence importance, symptoms, etiology, disease cycle and management of powdery mildew of pea.
3. Give the diagrammatic representation of disease cycle of the following diseases of pea

   i. Rust ii. Downy mildew

   iii. Anthracnose iv. Powdery mildew
4. Briefly describe the diagnostic symptoms and integrated management of any two of the following pea diseases :

   (i) Powdery mildew (ii) Anthracnose

   (iii) Wilt
5. Illustrate the powdery mildew of pea in the following headings :

   (i) Pathogen (ii) Symptoms

   (iii) Disease Cycle (iv) Disease Management
6. Describe in detail the most distinguishing symptoms, causal organism, disease cycle and management of anthracnose of pea.

## Section 'B'

### Short answer questions

1. How will you differentiate Erysiphe from Sphaerotheca ?
2. Why the diseases caused by Erysiphe are called powdery mildew?
3. Which fungal pathogen causes powdery mildew of pea? Give its systematic position.
4. Give the characteristic of the oospore of *Peronospora pisi.*
5. What are cleistothecia.

## Section 'C'

### Very short answer questions

(i) Suggest some disease resistant varieties of pea against with disease.

(ii) Mention the name the resting structures by the following pathogens :

  (a) *Fusarium oxysporum* f. sp. *pisi*

  (b) *Colletotrichum pisi*

  (c) *Erysiphe pisi*

  *(d) Uromyces fabae*

(iii) List some important diseases and their causal organisms of pea.

(iv) Distinguish between downy mildew and powdery mildew of pea.

(v) Which type of disease is can be effectively managed by mixed cropping.

## Section 'D'

### Objective type questions

(a) Fill in the blanks with suitable word (s)

  (i) *Uromyces pisi* is an.......................... .rust.

  (ii) Anthracnose of pea is caused by .........................

  (iii) Numerous black colored and thick walled ----- are produced in *Erysiphe polygoni*.

  (iv) Chlamydospore is the resting spore of the fungus ......

  (v) Foliar spray with ------------------ fungicide control powdery mildew of pea.

(b) State whether the following statements are True or False.

(i) *Erysiphe polygoni* ia an obligate parasite.

(ii) Foliar spray with Metalaxyl + Mancozeb @ 0.2% manage downy mildew of pea.

(iii) White floury patches symptom appear on both side of the leaves of pea in powdery mildew disease.

(iv) The fungus *Fusarium oxysporum* f.sp. *pisi* produces three types of conidia like macroconidia, microconidia and chlamydospores

(v) Wilt disease of pea can be managed by soil application of Trichoderma .

(vi) The fungus *Peronospora pisi* has coenocytic mycelium.

(vii) The fungus *Peronospora pisi* is facultative parasite

**Fig. 1:** Powdery mildew

**Fig. 2:** Downy mildew

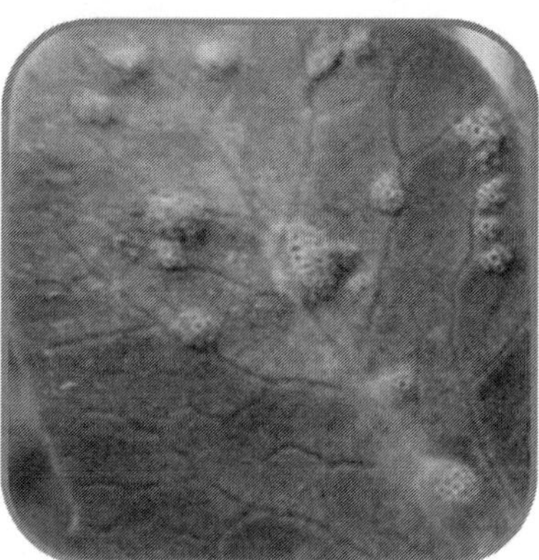

**Fig. 3:** Rust

**Plate 8:** Photograph showing symptoms of major diseases of pea *(See colour version on page 206)*

# DISEASES OF HORTICULTURAL CROPS

# 9

# Diseases of Mango and Their Management

| Sl.No | Name of Disease | Causal organism |
|---|---|---|
| 1 | Anthracnose | *Colletotrichum gloeosporioides* Ston, Spaull and Schrenk |
| 2 | Powdery mildew | *Oidium mangiferae* Berthet |
| 3 | Mango malformation | *Fusarium mangiferae* Wollenweber & Reinking |
| 4 | Bacterial blight | *Xanthomonas campestris* pv. *mangiferae* |
| 4 | Stem end rot | *Botrydiplodia theobromae* |
| 5 | Red rust | *Cephaleurus mycoides* |

## Anthracnose

***Colletotrichum gloeosporioides (Glomerella cingulata)***

### Diagnostic Symptoms

- Disease produces symptoms like leaf spot, blossom blight, withertip, twig blight and fruit rot .
- Symptoms on leaves show as gray to brown spots with dark margins and a yellow halo. The spots later enlarge and coaleasce to form sizable necrotic areas.
- Tender shoots and foliage are easily affected which ultimately cause die back of young branches.
- Blossom blight may vary in severity from slight to a heavy infection of the panicles .
- Black spots develop on panicles as well as on fruits. Severe infection destroys the entire inflorescence resulting in no setting of fruits.
- Young infected fruits develop black spots, shrivel and drop off.
- Fruits infected at mature stage carry the pathogen into storage and cause considerable loss during storage, transit and marketing (Fig. 2).

## Etiology

## Scientific Classification

Kingdom : Fungi

Division : Eumycota

Sub- division : Deuteromycotina

Class : Coelomycetes

Order : Melanconiales

Family : Melanconiaceae

Genus : *Colletotrichum*

Species : *gloeosporioides*

- *C. gloeosporioides* represents the anamorphic state of ascomycetous fungus *Glomerella cingulata*(Telemorphic state).
- Mycelium :septate and slightly dark coloured
- Hyphae: accumulate below the host cuticle and develop fruiting bodies called acervuli.
- Acervuli: consists of numerous, closely packed , single celled, colorless conidiophores that produce conidia on their tips.It may also posses sterile hair like structures called 'setae'.
- Conidia : broadly oval to oblong with obtuse ends, non septate, some times posses oil globules, measure 12-16x4-6μm.
- Setae : common on twigs but not on fruits.

## Disease Cycle

- Pathogen survives in plant debris like twigs and leaves which fall on the ground .
- Primary infection caused by condia developed on disesed plant parts on ground , are disseminated by air , brought on to the susceptible part of plant, germinate in favourable conditions.
- Secondary infection is mainly due to transfer of conidia through rain splash or wind driven rain water.

## Favourable Conditions

- Optimum temperature -25°C
- Relative Humidity -95-97% for 12 hours

## Integrated Management

- Plant resistant varieties and use healthy seedlings. Leave sufficient space between plants.
- Prune trees yearly to enhance ventilation.
- Implement good drainage methods.
- Diseased leaves, twigs, gall midge infected leaves and fruits, should be collected and burnt.
- Cover the fruits on tree, 15 days prior to harvest with news or brown paper bags.
- Foliar sprayof Azoxystrobin 23% SC (0.1%) or Copper oxy-chloride 50% WG (0.25%) at 15 days intervals until harvest.
- Foliar spray of *Pseudomonas fluorescens* @ 5g/liter of water at 21 days interval commencing from October on flower branches.5-7 sprays one to be given on flowers and bunches.
- Treat the fruit with hot water, (50-55°C) for 15 minutes or Thiobendazole (1000ppm) for 5 minutes before storage.
- Store fruits in a well ventilated environment.

## Powdery Mildew

### *Oidium mangiferae (Acrosporum mangiferae)*

## Diagnostic Symptoms

- Pathogen parasitizes young tissues of all parts of the inflorescence, leaves and fruits.
- Initially white superficial powdery fungal growth appear on leaves, stalks of panicles, flowers and young fruits.
- Young leaves are attacked on both the sides but it is more evident on the upper surface. Often these patches coalesce and occupy larger areas turning into purplish brown in colour.
- Affected flowers and fruits drop pre-maturely reducing the crop load considerably or might even prevent the fruit set (Figure -1).

## Etiology

### Scientific Classification

Kingdom : Fungi

Division : Eumycota

Class : Leotiomycetes

Order : Erysiphales

Family : Erysiphaceae

Genus : *Oidium*

Species : *mangifeare*

- Pathogen is an ectoparasitic biotroph and oogamous type.
- Hyphae are superficial, septate, hyaline, and ramify over the host surface forming white dense coating.
- Conidiophores short, hyaline and conidia single celled –barrel shaped, produced in chain.

### Disease Cycle

- Pathogen overwinters in dormant bud or on old leaves.
- Resting spores are source of primary infection.
- Secondary spread of disease can occur thorugh air borne conidia produced in these infections.
- Spores blown wind from infected areas readily adhere to hairy, unopened flowers near tip of the inflorescence and germinate in five to seven hours and causes infection.

### Favourable conditions

- Optimum temperature- 15-30$^0$C.
- Relative humidity-60-90 %.
- Rains or mists accompanied by cooler nights during flowering.
- Most susceptible stage: Full bloom and fruit set at pea size stage.

### Integrated Management

- Use resistant/tolerant varieties likeNeelum, Zardalu, Banglora, Torapari-khurd and Janardhan pasand.
- Foliar sprays with wettable sulphur @ 0.3% or Karathane@ 0.1% or Azoxystrobin 23% SC @ 0.1%

Hexaconazole 5% EC @ 0.1% or Triadimefon 25% WP @ 0.1% depending on the size of tree.

- Spraying at full bloom needs to be avoided.
- Grow mango on dry and well ventilated areas.
- Intercrop with other non host tree species.
- Prune diseased leaves and malformed panicles harbouring the pathogen to reduce primary inoculums load.

## Mango malformation

***Fusarium moliliforme*** *var.* ***subglutinans***

### Diagnostic Symptoms

- Two types of symptoms are generally produced;floral malformation and vegetative malformation.

- **Vegetative malformation**
  - Pronounced in young seedlings.
  - Affected seedlings develop vegetative growths which are abnormal growth, swollen and have very short internodes, giving it a bunch like an appearance on the shoot apex.
  - Seedling remain stunted and eventually die.
- **Floral malformation**
  - Flower buds are transformed into vegetative buds and a large number of small leaves and stems, which are characterized by appreciably reduced internodes and give an appearance of witches- broom.
  - Flower buds seldom open and remain dull green(Figure -4).

### Etiology

- Physiological disorder, eriophyid mites , fungal infection, virus, herbicides and other toxic compounds aids in the production of pathogen.
- Deficiency of Iron, Zinc and Copper can also cause the malformation.

### Disease Cycle

- Disease is mainly spread via infected plant material.
- Wound by mango bud mite lead to facilitate primary infection.
- Secondary infection is either by air borne conidia or by conidia carried by eriophid mite *Aceria mangiferae*

## Favourable Conditions

- Optimm temperature- 10-15°C.
- More disease in young than in old palnts.

## Integrated Management

- Use of disease free planting material.
- Monitor orchard regularly for any signs of deformed plants parts.
- Plough the field before planting to destroy existing weeds in the field.
- Deep summer ploughing.
- Irrigate the orchards as and when required.
- Provide proper shade, irrigation & drainage.
- Floral malformed panicles/ vegetative malformed shoots should be pruned and burnt.
- Application of NAA (200 ppm) in the first week of October (Before bud differentiation stages) followed by deblossoming in the late December or January.
- Spray Zinc, Boron and Copper before bloom and after fruit harvesting.

## Bacterial blight

***Xanthomanas mangiferae- indicae***

## Diagnostic Symptoms

- Also known as bacterial spot, leaf spot, black spot and mango blight.
- Symptoms appear on all the above plant parts.
- On the leaves initially slightly water soaked yellowish, translucent irregular spots are formed towards leaf tip which enlarge soon and become dark brown with a yellow halo. These dark spots are limited by veins and become angular resulting in cankerous raised lesions. Spots appear all over the lamina.Under high humid conditions these spots fuse, leaves turn yellow and drop down prematurely.
- On branches, green twigs, and stem, the fresh dark lesions are water soaked, which further become raised and dark brown having longitudinal fissures of infection. The vascular tissues beneath these lesions are filled with gum which oozes out giving a sticky appearance.
- When fruits are attacked, water soaked lesion develop and turn dark brown to black and gradually turn into cankers. Sometimes crack appear on the skin of infected fruits releasing gummy ooze containing bacterial cells. The severly affected fruits are shed prematurely (Fig. 3) .

## Etiology

### Scientific Classification

Kingdom : Prokaryotae

Division : Gracilicutes

Class : Proteobacteria

Family : Psudomonadaceae

Genus : *Xanthomonas*

Species : *mangiferae- indicae*

- Gram-ve, rod, non-spore former.
- Motile with single polar flagellum.
- Colonies are smooth, butyrous, circular, glistening, convex with entire margin, pale to light yellow in color .
- Yellowish color of the colonies is due to production of characteristic water insoluble, non-diffusible pigment.

### Disease Cycle

- Bacterium survives on diseased plant parts where it multiplies at early stages.
- Primary infection is through infected leaves, branches and fruits.
- Secondary infection occurs by wind driven rain and rain splash.
- Pathogen enters mango leaves through stomata and fruits through natural opening and wounds.

### Favourable conditions

- High temperature-32$^0$C
- High humidity

### Integrated Management

- Use healthy planting material.
- Orhard sanitation and seedling certification.
- Grow resistant variety *viz.*, Bombay green, Jahagir, Fazali and Suvarnrekha and moderately resistant varieties *viz.*, Langra, Dashehari, Chausa, Bombay, Zardalu, Gulabkhas, Kesar.
- Foliar spray of 0.02% Streptomycin followed by Copper oxychloride (0.3%).

- Ensure good ventilation of the trees.
- Avoid mechanical injury to the mango trees during field work.
- Protect them from strong winds and heavy rains with wind breaks.
- Disinfect working tools and equipments.

## MODEL QUESTION PAPER

### Section 'A'

### Long answer questions

1. Describe in detail the occurrence importance, symptoms, etiology, disease cycle and management of powdery mildew of mango.
2. Give the diagrammatic representation of disease cycle of the following diseases of mango
   i. Mango malformation  ii. Bacterial blight
   iii. Anthracnose  iv. Powdery mildew
3. Briefly describe the diagnostic symptoms and integrated management of any two of the following mango diseases :
   (i) Powdery mildew  (ii) Anthracnose
   (iii) Bacterial blight
4. Explain the mango malformation in the following headings:
   (i) Pathogen  (ii) Symptoms
   (ii) Disease cycle  (iv) Disease management
5. Describe in detail the most distinguishing symptoms, causal organism, disease cycle and integrated management of anthracnose of mango.

### Section 'B'

### Short answer questions

1. How will you differentiate vegetative malformation from floral malformation.
2. Write down the etiology of mango malformation.
3. Which fungal pathogen causes powdery mildew of mango? Give its systematic position.
4. Write down the management practices of bacterial blight of mango.
5. Describe the causal organisms and mode of penetration of powdery mildew of mango.

## Section 'C'

### Very short answer questions

(i) Suggest some chemicals for management of powdery mildew of mango.

(ii) Write down the etiology of powdery mildew disease of mango

(iii) List some important post harvest diseases of mango and their causal organisms.

(iv) Mention the favorable condition for development of powdery mildew of mango.

(v) Which type of disease can be effectively managed by application of bordeaux paste.

## Section 'D'

### Objective type questions

Fill in the blanks with suitable word (s)

i. Red rust of mango is caused by …………………….. ..

ii. Anthracnose of mango is caused by ……………

*iii.* *Colletotrichum gloeosporioides* represents the anamorphic state of ascomycetous fungus ……

iv. Foliar spray with karathane fungicide control --------------- disease of mango.

v. Acervuli of *Colletotrichum gloeosporioides* posses sterile hair like structures called ----------------.

vi. *Xanthomonas mangiferae- indicae* is a gram -------------- rod and non-spore forming bacteria.

### State whether the following statements are True or False

*i.* *Oidium mangiferae* ia an obligate parasite.

ii. Secondary spread of powdery mildew of mango can occur thorugh air borne conidia .

iii. White floury patches symptom appear on both side of the leaves of mango in powdery mildew disease.

iv. Yellowish color of the colonies is due to production of characteristic water insoluble, non-diffusible pigment.

v. The hyphae of *Colletotrichum gloeosporioides* accumulate below the host cuticle and develop fruiting bodies called acervuli.

**Fig. 1:** Powdery mildew

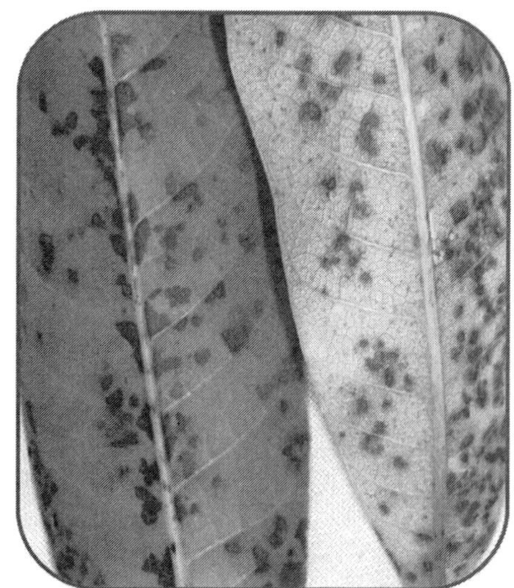

**Fig. 2:** Anthracnose

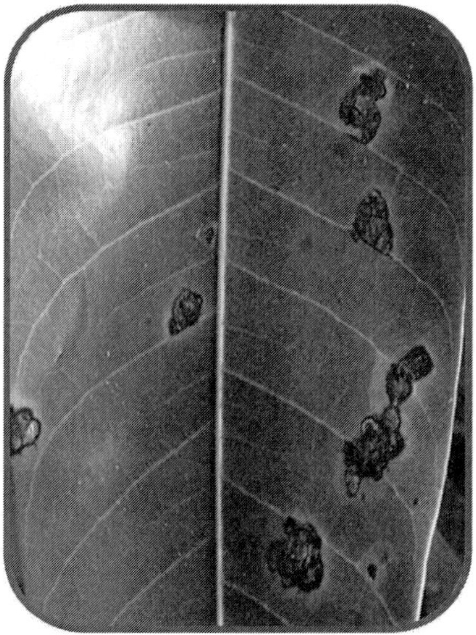

**Fig. 3:** Bacterial black spot

**Fig. 4:** Flower malformation

**Plate 9:** Photograph showing symptoms of major diseases of mango *(See Colour version on page 207)*

# 10

# Diseases of Citrus and Their Management

| Sl.No | Name of Disease | Causal organism |
|---|---|---|
| 1 | Gummosis | *Phytophthora* spp |
| 2 | Canker | *Xanthomonas axonopodis* pv. *citri* (Hasse) |
| 3 | Scab/Verucosis | *Elsinoe fawcett* |
| 4 | Tristeza or quick decline | *Citrus tristeza virus* |
| 5 | Greening | *Liberobacterium asiaticum* |
| 6 | Exocortis or scaly butt | Viroid |

## Gummosis

## *Phytophthora* spp

### Diagnostic Symptoms

- Dark water soaked large patches appear on the basal portions of the stem near the ground level.
- Bark in such parts dries, shrinks and cracks and shreds in lengthwise vertical strips.
- Later profuse exudation of gum from the bark of the trunk occurs.
- Considerable amount of gum formation in sweet oranges may be observed, but relatively little in grapefruit (Figure -1).

### Etiology

### Scientific Classification

Kingdom : Fungi

Division : Eumycota

Class : Oomycetes

Order : Peronosporales

Family : Pythiaceae

Genus : *Phytophthora*

Species : *palmivora/parasitica*

- *P. palmivora* is the main cause of the disease in southern parts of our country although *P. parasitica* is also common in Karnataka.*P. parasitica* is found to cause the disease in Assam

| Character | *P. palmivora* | *P. parasitica* |
|---|---|---|
| Hyphae | Intercellular with haustoria and often swollen at regular intervals | Tough and irrregular |
| Sporangiophores | Simple or branched | Slender, irregular or sympodically branched |
| Sporangia | Obpyriform and always terminal | Lateral or intercalary, broadly ovoid, obpyriform to spherical,papillate |
| Oospores | Spherical, thickwalled and produce secondary sporangium on germination | Apleurotic |

## Disease Cycle

- Pathogen survives in soil and plant debris.
- Primary source of infection is zoospores splashed on to the tree trunk.
- Secondary infection through spores that may be transported in rain or irrigation to the roots of the trees.
- They germinate and enter the root tip resulting in rot of the entire root let, later extending to the rest of the root.

## Favourable Conditions

- Optimum temperature -25-28$^0$C.
- Soil pH-5.4-7.5.
- Prolonged contact of trunk with water as in flood irrigation, water logged areas and heavy soils.

## Integrated Management

- Remove diseased bark and a buffer strip of healthy, light brown to greenish bark around the margins of the infection.
- Selection of proper site with adequate drainage.
- Use of resistant rootstocks.
- Provision of an inner ring about 45 cm around the tree trunk to prevent moist soil.
- Avoid irrigation water from coming in direct contact with the trunk.
- Avoid injuries to crown roots or base of stem during cultural operations.
- Painting Bordeaux paste to a height of about 60 cm above the ground level at least once a year.
- Scrape the diseased portion with a sharp knife.

- Protect the cut surface with Bordeaux paste followed by spraying of 0.3% Fosetyl-AL reduces the spread.
- Soil drenching with 0.2% Metalaxyl and 0.5% *Trichoderma viride* commercial formulation or Aureofungin 46.15% w/v. SP @ 300 (1%) gm /ml/acre.

## 2. Canker

### *Xanthomonas campestris* pv *citri*

### Diagnostic Symptoms

- Cankers are slightly sunken areas surrounded by irregular cracks and raised margin.
- Initially, disease appears as minute water soaked round, yellow spots which enlarge slightly and turn brown, eruptive and corky.
- These pustules are surrounded by a characteristic yellow halo.
- Canker lesions on the fruit do not possess the yellow halo as on leaves.
- Several lesions on fruit may coalesce to form larger canker.
- Due to severe infections, there may be defoliation, and twig and stem may show die-back symptoms.
- Canker lesions on fruits reduce market value.
- Plants become stunted and fruit yields reduced considerably (Figure -2).

### Etiology

### Scientific Classification

Kingdom : Prokaryotae

Division : Gracilicutes

Class : Proteobacteria

Family : Psudomonadaceae

Genus : *Xanthomonas*

Species : *campestris* pv. *citri*

- Pathogen is gram negative, non spore forming and aerobic bacteria.
- Rod shaped.
- Size: 1.5-2.0 x0.5-0.75 μm.
- Forms chains and capsules.
- Motile by one polar flagellum.

- Colony: circular, straw yellow colored, slightly raised, smooth and shining.
- There are at least three types of citrus canker, caused by different strains. Cancrosis A, Cancrosis B &Cancrosis C.
- Cancrosis A (The Asiatic type) ,caused by A-strain, is the most severe form of the disease. It is found in India .

## Disease Cycle

- Pathogen survives and multiplies in infected leaves, twigs and fruits.
- It may also survive in crevices in bark tissues of citrus trees.
- Infection occurs when wind blown rain carries inoculums to uninfected plants.
- Bacteria enter the leaf through stomata and lenticels, along with water and colonize the intercellular spaces.
- Infection also occurs through wounds caused by Citrus leaf miner larvae(*Phyllocnistis citrella*).
- Infection occurs on young leaves, twigs and the fruits.

## Favourable Conditions

- Favourable temperature- 20-30°C.
- Free moisture for 20 minutes.

## Integrated Management

- Select seedlings free from canker for planting in main field.
- Prune out and burn all canker infected twigs before monsoon.
- Maintain proper aeration by training and pruning.
- Sterilize tools and equipment between uses to prevent the spread of the disease.
- Do not work in the field when foliage is wet.
- Clean boot, clothes thoroughly when working between different orchards.
- Use windbreaks between fields to avoid propagation.
- Monitor the trees having signs of the disease.
- Foliar spray with Streptocycline (1g) + Copper oxy chloride (30g) in 10 litres of water at fortnightly intervals .
- SprayStreptocycline 50 to100 ppm solution repeatedly at an interval of 15 to 20 days after the appearance of new growth.
- Cover the foliage and young fruits fully.

## MODEL QUESTION PAPER

### Section 'A'

### Long answer questions

1. Explain in detail the occurrence, importance, symptoms, etiology, disease cycle and management of citrus canker .
2. Write down the favorable conditions for development of the following diseases of citrus
   - i. Citrus canker
   - ii. Gummosis
   - iii. Tristiza
   - iv. Greening
3. Briefly describe the diagnostic symptoms and integrated management of any two of the following citrus diseases:
   - (i) Greening
   - (ii) Tristeza
   - (iii) Canker
4. Describe the gummosis in the following headings:
   - (i) Pathogen
   - (ii) Symptoms
   - (ii) Disease cycle
   - (iv) Disease management

### Section 'B'

### Short answer questions

1. Which species of citrus has been found to be the most susceptible to gummosis? Which pathogens cause this disease?
2. Write down the etiology of gummosis.
3. Which bacterial pathogen causes citrus canker? Give its systematic position.
4. Write down the integrated management practices of gummosis.
5. Describe the causal organisms and mode of penetration of citrus canker.

### Section 'C'

### Objective type questions

Fill in the blanks with suitable word (s)

i. Citrus canker is caused by ……………………….. ..

ii. Gummosis of citrus is caused by ……………

iii. Primary source of infection in gummosis of citrus is --------------------

iv. Metalaxyl fungicide used for control of --------------- disease of citrus.

v. The color of colony of *Xanthomonas campestris* pv *citri* is --------------.

vi. *Xanthomonas campestris* pv *citri* is a gram -------------- bacteria.

State whether the following statements are True or False.

i. *Xanthomonas campestris* pv *citri* is a rod shaped bacteria
ii. The sporangia of *Phytophthora palmivora* is obpyriform and terminal.
iii. The oospore of *Phytophthora palmivora* is spherical, thick walled and produce secondary sporangium on germination.
iv. The colony of *Xanthomonas campestris* pv *citri*: is circular, straw yellow colored, slightly raised, smooth and shining.
v. The hyphae of *Phytophthora palmivora* is intercellular with haustoria and often swollen at regular intervals

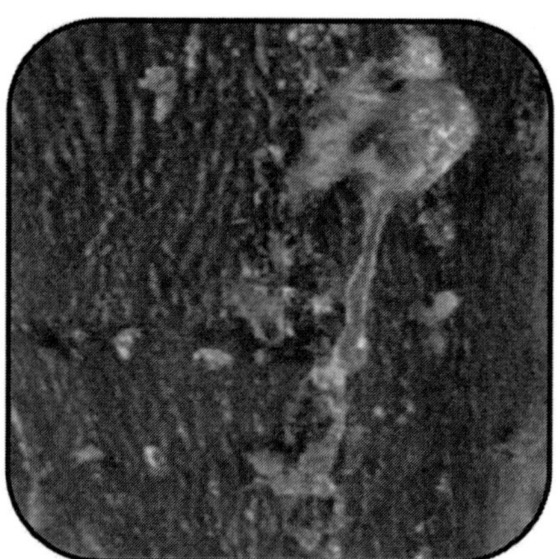

**Fig. 1:** Gummosis

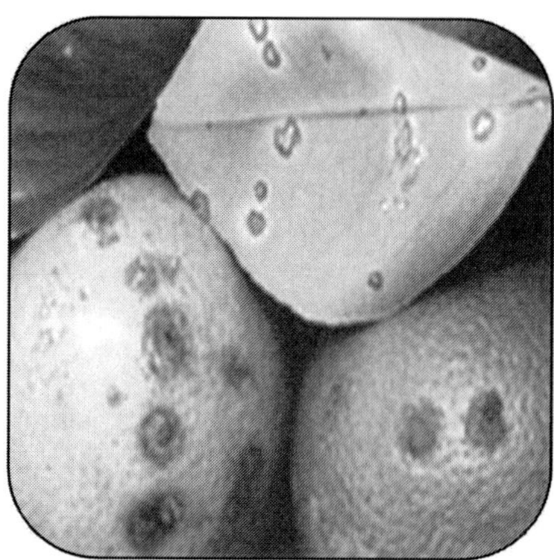

**Fig. 2:** Canker

**Plate 10:** Photograph showing symptoms of major diseases of citrus *(See colour version on page 208)*

# 11

# Diseases of Grape and Their Management

| Sl.No | Name of Disease | Causal organism |
|---|---|---|
| 1 | Downy mildew | *Plasmopara viticola* (Berk. & Curt.) Berl.& de Toni |
| 2 | Powdery mildew | *Uncinula necator* (Schw.) Burr. |
| 3 | Anthracnose | *Elsinoe ampelina* (de Bary) Shear |

## Downy Mildew

### *Plasmopara viticola*

- Downy mildew of grapes was catastrophic in France in 1878.
- Millardet observed that a farmer had sprinkled a mixture of copper sulphate and lime on the vines along the rode side to prevent pilferage of grapes.
- Millardet after extensive experimentation , developed Bordeaux mixture which proved a boon and controlled the disease with spectacular success.

## Diagnostic Symptoms

- Pathogen can attack all green parts of the vine.
- Infected leaves develop pale yellow-green lesions which gradually turn brown.
- Severely infected leaves often drop prematurely.
- Infected petioles, tendrils, and shoots often curl, develop a shepherd's crook, and eventually turn brown and die.
- Young berries are highly susceptible to infection and are often covered with white fruiting structures of the pathogen.
- Infected older berries of white cultivars may turn dull gray green, whereas those of black cultivars turn pinkish red.
- No cracking of the skin of the berries.
- Symptoms of this disease are frequently confused with those of powdery mildew Fig. 2

## Etiology

### Scientific Classification

Kingdom : Fungi

Division : Eumycota

Class : Oomycetes

Order : Peronosporales

Family : Peronosporaceae

Genus : *Plasmopara*

Species : *viticola*

- Pathogen is biotroph.
- Mycelium: intercellular with spherical haustoria, coenocytic, thin walled and hyaline.
- Sporangiophores: arise from hyphae in the sub stomatal spaces. It branched at right angle to the main axis and at regular intervals. Secondary branches arise from lower branches.
- Sporangia: thin walled, oval or lemon shaped germinate either by forming zoospores, or behave as conidia and germinate by germ tube.
- Zoospores: egg shaped, biflagellate and 7 – 9 micron meter.
- Oospores: thick walled with a rough epispore and are produced mostly in tissues adjacent to the midrib.

### Disease Cycle

- Pathogen overwinters mainly as oospores in the debris in the soil, serve as the primary inoculum.
- They germinate and form numerous biflagellate zoospores.
- Primary infection is by zoospores transported by wind or water to the wet lower leaves near the soil line.
- They encyst and form a germ tube that penetrates through stomata or the epidermis.
- Infection becomes systemic and symptom appear after 10-15 days of infection.
- Secondary infection takes place by the motile zoospores by splashing rain.

- As the close of the season, oospores are formed, which fainally reach the soil.

**Favourable conditions**

- Optimum temperature : 20-22°C
- Relative humidity : 80-100 per cent
- Wet winter followed by a wet spring and a warm summer with intermittent rains.

**Integrated Management**

- Deep ploughing of fields during summer.
- Make sure soils are well drained.
- Field sanitation.
- Remove weed and plant residues from the field.
- Keep tools and equipment clean.
- Aply balanced fertilizer
- Destroy the alternate host plants as well as crop debris.
- Vine should be kept high above ground to allow circulation of air by proper spacing
- Pruning (April- May & September –October) and burning of infected twigs.
- Grow resistant varieties like Amber Queen, Cardinal, Champa, Champion, Dogridge and Red Sultana
- Foliar spray with Copper oxychloride 50% WG@ 0.3% or Cymoxanil 50 % WP@ 0.25% or 0.2% or Propineb70%WP @ 0.30% or Cymoxanil8%+ Mancozeb64%WP @ 0.2% or Metalaxyl M 4%+ Mancozeb 64% WP @ 0.25% or Metalaxyl8%+ Mancozeb 64% WP@ 0.2%, if required repeat after 15 days.

**Powdery mildew**

*Uncinula necator*

Diagnostic Symptoms

- Pathogen can infect all green tissues of the grapevine.
- Diseased leaves appear whitish gray, dusty, or have a powdery white appearance.
- Petioles, cluster stems, and green shoots often look distorted or stunted.
- Berries can be infected until their sugar content reaches about 8%.

- If infected when young, the epidermis of the berry can split and the berries dry up or rot.
- When older berries are infected, a netlike pattern often develops on the surface of the berry (Fig-1).

## Etiology

### Scientific Classification

Kingdom : Fungi

Division : Eumycota

Class : Leotiomycetes

Order : Erysiphales

Family : Erysiphaceae

Genus : *Uncinula*

*Species : necator*

- Pathogen is oidium type and obligate parasite.
- White growth consists of mycelium, conidiophores and conidia.
- Mycelium: external, septate and hyaline.
- Conidiophores: short and arise from external mycelium.
- Conidia : produced in chain, single celled, hyaline and barrel shaped.
- Cleistothecia: black, depressed, contains 4-8 asci.
- Ascus: contains 4-6 oval ascospores.

### Disease Cycle

- Pathogen overwinters as mycelium in the dormant buds and as cleistotheica in the cracks in the bark.Besides, the collateral hosts and volunteer plants provide conidia as the primary mycelium.
- Infected buds give rise to systematically infected shoots.
- Cleistothecia, when wet, discharge the ascospores, which infect the leaves.
- Ascospores and conidia germinate and push haustoria into the epidermal cells from the appresorium.
- Mycelium grows externally and forms conidia in 7 to 10 days after infection.
- Secondary infections can occur through spores produced, released when conditions are favorable for growth of the fungus in spring and cause new infections.

### Favourable conditions

- Cool dry weather.
- Optimum temperature – 20-25 $^0$C
- Enough relative humidity.

### Integrated Management

- Deep ploughing of fields during summer .
- Use tolerant varieties if available.
- Keep a wide distance between vines to allow for good air circulation.
- Alternativley chose pruning practices that favor an open canopty.
- Monitor the field regularly for signs of the disease.
- Use nitrogen fertilizers with caution to avoid excessive vegetative growth.
- Field sanitation.
- Destroy the alternate host plants as well as crop debris.
- Adopt ecological engineering by growing the attractant, repellent, and trap crops around the field bunds.
- Grow resistant varieties like ChholthRed ,Skibba Red, Chholth white
- Foliar spray with Wettable Sulphur @0.3% or Calixin. @ 0.1% or Flusilazole 40% EC@ 0.01% or Myclobutanil 10% WP @ 0.04% or Sulphur 55.16 % SC @ 0.30% or Triadimefon 25% WP@ 0.010% , sprayed at 15 days interval .

### Anthracnose /Bird's Eye Spot

*Gloeosporium ampelophagum* **(***Elsinoe ampelina)*

### Diagnostic Symptoms

- Pathogen attacks shoots, tendrils, petioles, leaf veins and berry.
- Symptoms appears first as dark red spots on the berry. These spots are circular, sunken, ashy-gray and in late stages these spots are surrounded by a dark margin which gives it the "bird s-eye rot" appearance.
- Spots vary in size from ¼ inch in diameter to about half the fruit.
- Numerous spots sometimes occur on the young shoots.
- These spots may unite and girdle the stem, causing death of the tips.
- Spots on petioles and leaves cause them to curl or become distorted (Fig-3).

## Etiology

### Scientific Classification

Kingdom : Fungi

Division : Eumycota

Phyllum : Deuteromycotina

Class : Coelomycetes

Order : Melancoliales

Family : Melanconiaceae

Genus : *Gloeosporium*

Species : *ampelophagum*

- Perfect stage : *Elsinoe alpelina*
- Hyphae: well developed, septate and branched.They accumulate in subcuticular region of infected plant and develop acervuli.
- Acervuli: single celled , colorless , delicate and small conidiophores are formed in large number in each acervulus.
- Conidia: produce on tips of conidiophores, unicellular, hyaline, slightly narrower in the middle than at the ends, measure 5-6 x2.5-3.5µm.

### Disease Cycle

- Pathogen survives in infected plant debris and act as a source of inoculums.
- Primary infection is through spores produced by disease vines.
- Secondary infection by wind borne conidia.

### Favourable Conditions

- Optimum temperature-24-26$^{0}$C.
- Leaf wetness period-3 hrs.
- Humid and wet weather

### Integrated Management

- Choose sites with suitable sun exposure and good air circulation.
- Use of resistant varieties such as– Bangalore Blue, Beauty Seedless, Bharat Early, Golden queen, Large white.
- Make sure to keep a wide distance between vines.

- Remove any wild grapes near the vineyard.
- Monitor the vines and remove fruit or plant parts that show sign of the disease.
- Prune the vines in early winter during dormancy.
- Remove plant residues from the vineyard plow field and burry.
- Foliar spray with Aureofungin 46.15%w/v. SP@ 0.005% or Carbendazim 50%WP@ 0.1% or Chlorothalonil 75% WP@ 0.2% or Kitazin 48% EC @ 0.20% or Mancozeb75% WP @ 0.25%, if required repeat after 15 days
- Removal of infected twigs.

## MODEL QUESTION PAPER

### Section 'A'

### Long answer questions

1. Explain in detail the occurrence, importance, symptoms, etiology, disease cycle and management of downy mildew.
2. Briefly describe the diagnostic symptoms and integrated management of any two of the following grape diseases:

   (i) Powdery mildew (ii) Downy mildew

   (iii) Anthracnose
3. Describe the powdery mildew of grapes in the following headings:

   (i) Pathogen (ii) Symptoms

   (ii) Disease cycle (iv) Disease management\
4. Describe in detail the most distinguishing symptoms, causal organism, disease cycle and integrated management of anthracnose of grapes.
5. Give the diagrammatic representation of the disease cycle of the following diseases of grapes

   (i) Powdery mildew (ii) Downy mildew

   (iii) Anthracnose

### Section 'B'

### Short answer questions

1. How will you differentiate downy mildew of grape from powdery mildew of grape.
2. Write down the etiology of powdery mildew of grape.

3. Which fungal pathogen causes downy mildew of grape? Give its systematic position.
4. Write down the integrated management practices of anthracnose of grape.
5. Describe the causal organisms and mode of penetration of powdery mildew of grape.
6. What is Bordeaux mixture?
7. What is the mode of branching of the sporangiophore of *Plasmopara viticola?*
8. Why the downy mildew disease of grape is historically important ?
9. Suggest some chemicals for management of downy mildew of grape.
10. Which type of disease can be effectively managed by mixed cropping.

**Section 'D'**

## Objective type questions

Fill in the blanks with suitable word (s)

i. Downy mildew of grape is caused by ......................... ..
ii. Anthracnose of grape is caused by ..............
iii. The telemorphic stage of anthracnose of grape is ......
iv. Foliar spray with sulfur fungicide control ---------------- disease of grape.
v. ---------------------------made the accidental discovery of Bordeaux mixture.
vi. The fungus *Gloesporium ampelophagum* belongs to family-----------------

**State whether the following statements are True or False.**

*i.* *Plasmopara viticola* ia an obligate parasite.
*ii.* *Plasmopara viticola* belongs to class oomycetes.
*iii.* *Plasmopara viticola* perenates mainly by means of spores.
iv. White floury patches symptom appear on both side of the leaves of grape in powdery mildew disease.
v. The acervuli of *Gloeosporium* do not produce setae.
vi. The hyphae of *Colletotrichumg gloeosporioides* accumulate below the host cuticle and develop fruiting bodies called acervuli.
vii. Dry areas remain usually free of downy mildew of grape.

**Fig. 1:** Powdery mildew

**Fig. 2:** Downy mildew

**Fig. 3:** Anthracnose

**Plate 11:** Photograph showing symptoms of major diseases of grape *(See colour version on page 208)*

# 12

# Diseases of Apple and Their Management

| Sl No | Name of Diseases | Causal Organism |
|---|---|---|
| 1 | Scab | *Venturia inaequalis*(Cooke) Wint |
| 2 | Powdery mildew | *Podosphaera leucotricha*(Ell. And Ev.) |
| 3 | Fire blight | *Erwinia amylovora* (Burrill) Winslow et al |
| 4 | Crown gall | *Agrobacterium tumefaciens* (E.F.Sm.&Towns) Conn |

## Scab

***Ventruia inaequalis (Spilocaea pomi* Fr.*)***

### Diagnostic Symptoms

- Symptoms generally noticed on leaves and fruits.
- Affected leaves become twisted or puckered and have black, circular spots on their upper surface.
- Spots are velvety and may coalesce to cover the whole leaf surface. Severely affected leaves may turn yellow and drop.
- Scab can also infect flower, stems and cause flowers to drop.
- Lesions later become sunken and brown and may have spores around their margins.
- Infected fruit become distorted and may crack, allowing entry of secondary organisms(Figure-1)

### Etiology

### Scientific Classification

Kingdom : Fungi

Division : Ascomycota

Class : Loculoascomycetes

Order : Pleosporales

Family : Venturiaceae

Genus : *Venturia*

Species : *inaequalis*

- Pathogen has got two distinct stages- the imperfect stage(*Spilocaea pomi*) stage on living plant parts and perfect (*Venturia inaequalis*) on fallen dead leaves.
- Fungus produces ascocarps in a stroma in over wintered leaves in orchard floor.
- Pseudothecia are dark brown to black and spherical 90-150µm in diameter, having short beak and a distinct ostiole having single celled bristles at the apex.
- Asci: are cylindrical, fasciculate short stipulate produced in large number in the pseudotheicum which vary in size from 55-75x6-12 µm.
- Ascospores: measure 11-15x5-7 µm which are yellowish green to unequally bicelled, with upper cell smaller and wider than the lower one.
- Conidia are produced on conidiophores which arise from short erect, closely septate brown mycelium and are non septate or have a single septum. Conidia are produced on leaf surface and fruit lesions throughput the spring and summer and measure 12-12x6-9µm.
- Fungus is bipolar heterothallic.
- There are five races of the pathogen .Out of these two have been reported to occur in India.

## Disease cycle

- Pathogen survives through perithecia in the soil debris.
- Soil inoculums is the source of primary infection.
- Secondary infections occurs by conidia through rain splash or wind

## Favourable conditions

- High soil moisture.
- Temperature -25$^{0}$ C.
- Leaf wetness.

## Integrated Management

- Use resistant or tolerant varieties.
- Collect and burn of fallen leaves, shoots and fruits.
- Ensure pruning method that allow for more air circulation.
- Avoid getting foliage wet when watering.

- Resistant varieties: Ashwini, Ambika, Angeles, American pride, Surabhi.
- Sprays with wettable Difenoconazole @0.1% or Captan @0.25% or Carbendazim@0.1% at15 days interval.

## Powdery mildew

***Podosphaera leucotricha (Oidium farinosum)***

### Diagnostic Symptoms

- Symptoms produced on young shoots, leaves, blossoms and fruit.
- White or grey patches appears on the leaves.
- Leaf fold, twist and cracking may takes place.
- Buds get infected and may be killed.
- Both leaves and twigs may be covered with white powdery mass.
- Defoliation of leaves and deformation of fruits takes place. (Figure-2)

### Etiology

### Scientific Classification

Kingdom : Fungi

Division : Eumycota

Phyllum : Ascomycota

Class : Ascomycetes

Order : Erysiphales

Family : Erysiphaceae

Genus : *Oidium*

*Species : farinosum*

- Obligate parasite
- Pathogen consists of weft of white hyphal threads and draw nutrition through th haustoria.
- Conidia :produced in chain on conidiophores,hyaline clear, without color, measure 20-38 × 12 μm, and contain distinct fibrosin bodies.
- Cleistothecia: ascocarp of pathogen, globose, black, partly embedded in mycelium.Appendages are long and stiff or short and tortuous
- Asci: sac-like , enclosed in fruiting bodies ascocarps, contains a single ascus with eight ascospores, each of which is elliptical and measures 22-

36 x 12-15 µm. Ascocarps are densely grouped together, measure 75-96 µm in diameter and have apical and basal appendages.

**Disease Cycle**

- Pathogen survives in the form of resting mycelium or encapsulated haustoria in the dormant terminal and lateral shoot buds and in blossom buds produced and infect the previous growing season.
- Abundant sporulation from overwintering shoots and secondary lesions on young foliage leads to a rapid build up of inoculum.
- Secondary spread occurs through wind borne conidia.
- Cleistotheica are produced on heavily infected shoot and leaves in mid summer but they are not regarded as an important inoculums source because the ascoposres do not germinate readily.

**Favourable conditions**

- High relative humidity –greater than 70%.
- Temperature – 10 and 25°C.
- Youngest leaves are the most susceptible.

**Integrated Management**

- Plant resistant varieties.
- Plant crops with sufficient spacing to allow for good ventilation.
- Remove infect leaves when the first spots appear.
- Donot touch healthy plants after touching infected plants.
- Fertilize with a balanced nutrient supply.
- Remove plant residue after harvest.
- Foliar spray with Axoxystrobin + Difenoconazole (0.1%) or Boscalid +Pyraclostrobin (0.1%) or Thiophanate methyl 70% WP @ 0.1% or Calixin 0.1%, if required repeat after 15 days.
- Monitor fields regularly to assess the incidence of a disease.

**Fire blight**

***Erwinia amylovora* pv. *pyri***

**Diagnostic Symptoms**

- Term 'fire blight' sums up the symptoms as if scorched by fire. Primary infection occurs on flowers causing 'blossom blight', followed by the more serious 'shoot blight'.

- Infected blossoms become brown, later turning black and giving a burnt appearance.
- Flower become water soaked and brownish to black in color,may fall or remain hanging.
- Brwon black blotches appear on the leaves, blackening advances and leaves curl and hang.
- Water sprouts and terminal twigs wilt, die-back extends down the water sprout and twigs.
- Canker appear on branches which may girdle the branch and trunk to cause collar rot.
- Infected fruit becomes water soaked and brown to black in color. Brown ooze may be seen on the surface of infected parts under humid conditions.

## Etiology

### Scientific Classification

Kingdom : Bacteria

Phyllum : Protobacteria

Class : Gammaproteobacteria

Order : Enterobacteriales

Family : Enterobacteriaceae

Genus : *Erwinia*

Species : *amylovora* pv.*pyri*

- The pathogen is of historical importance. It has the distinction of being the first bacterium of plants shown to be a pathogen.
- The bacterium can be a plant pathogen was given by T.J.Burrill in 1878.
- *Erwinia amyovora* is also the first pathogen to be associated with an inscet vector.
- Rod shaped, filamentous bacterium.
- Gram negative.
- Size 1.1-3.0 µm x 0.5-1.2 µm.
- Motile by many peritrichous flagella, aerobic or facultative anaerobic, non sporulating.

- Colonies of virulent isolates are small, round and white with typical glistening shine.
- Bacterium requires nicotininc acid as a growth factor.

## Disease Cycle

- Pathogen survives primarily in cankers, blighted twigs and bud of symptomless twigs and serve as primary inoculums for infection next spring.
- During spring and early summer, infected parts reactivate and produce bacterial ooze consisting of millions of bacterial cells and easily transported to blossoms by insects such as flies, ants, and beetles.
- Bacteria can then be spread very efficiently from blossom to blossom by honey bees causes blossom blight.
- Once blossoms are infected, the bacteria can quickly spread and cause secondary infection into shoots and branches, which can be spread by rain, insects, and contaminated pruning tools.Secondary infections may continue to occur throughout the growing season.
- Secondary infections multiply the disease and finally result in infections of the tree trunk and the root stock.

## Favourable Conditions

- Temperature- 21 to 23$^{0}$C.
- Warm cloudy and rainy weather.

## Integrated Management

- Grow resistant varieties, if available.
- Monitor orchards regularly for signs of the disease.
- Prune out infected branches and burn them , preferably by the end of the winter.
- Donot apply excess nitrogen on the trees.
- Shave off the cankers and treat with Bordeaux paste.
- Foliar spray with Bordeaux mixture (0.1%) along with Streptocylin or Terramycin.
- Regulate soil fertility to reduce succulence of new growth.

## Crown Gall

### *Agrobacterium tumefaciens*

### Diagnostic Symptoms

- Small tumours or galls develop in the wound just below the soil surface. They enlarge rapidly with outer tissues becoming brown or black.
- Galls can also appear on roots, stem, branches, petioles and leaf veins.
- In early stages, tumours are more or leass spherical, white colored and quite soft.Later , the outer tissue becomes dark brown.
- Gall tissue is disorganized growth with an enlarged cambium and irregular vascular tissue. Movement of water and nutrients is cleverly impaired by galls.
- Movement of water and nutrients is severly impaired by galls.
- Attacked plants may remain stunted producing small chlorotic leaves. Main roots grow poorly and their yiled is reduced (Figure-3).

### Etiology

### Scientific Classification

Kingdom : Bacteria

Division : Gracilicutes

Class : Proteobacteria

Order : Rhizobiales

Family : Rhizoblaceae

Genus : *Agrobacterium*

Species : *tumefaciens*

- Shape: Rod shaped
- Staining: Gram negative
- Size: 0.6-1.0 μm x 1.5-3.0 μm
- Nature: Aerobic
- Donot produce endospores
- Flagella: 2-4, peritrichous flagella
- Plasmids: Pathogen contain one to several large plasmids composed of circular double stranded DNA.

- Ti-Plasmid: Plasmid carries the genes for gall or tumour induction and is called tumour inducing plasmid (Ti-Plasmid).
- Most important characteristic feature of pathogen is its ability to transfer part of Ti-plasmid (T-DNA) into host plant chromosome and transform normal plant cells to tumor cells.
- This has two principal physiological effects; it disrupt the metabolism of growth hormones in the tissues, brining about oncogenesis and it causes the synthesis of specific chemicals called opines. These opines can be utilized only by a bacteria that contain an appropriate Ti-plasmid and make it a 'genetic parasite'.
- High phytohormone level results from expression of three oncogenes transferred stably into the plant genome from *A. tumefaciens* : iaaM, iaaH and ipt.
- iaaM and iaaH oncogenes direct auxin biosynthesis and ipt oncogene cause cytokinin production.

## Disease Cycle

- Pathogen is a soil inhabitant and perennates in infested soil where it can live as a saprophyte for several years.
- When susceptible plants are raised at such sites , the pathogen reaches the crown region through the freshly developed wounds due to cultural practices, grafting or by insects.
- Once bacteria reaches inside the tissue, produce an enzyme that cuts the Ti-plasmid at specific sites releasing the segment of T-DNA form the Ri or Ti plasmid is integrated into a plant cell chromosome at various site and expressed.
- T-DNA expression results in overproduction of plant hormones and usually the synthesis of opines. Uncontrolled synthesis of hormones stimulates plant cells to divide , enlarge and form a tumor.
- With the increase in number and size of tumor cells,tumours exert pressure on surrounding and underlying normal tissues, which may be distorted xylem vessels and ultimately impaired the water supply to upper parts of plants .
- Due to invasion of tumours by secondary microbes decay of peripheral cells takes place.
- The bacterium from these galls may escape to contaminate the surrounding soil of healthy roots.

- The pathogen is then disseminated to new planting sites and healthy plants by splashing rain, irrigation water, tools, wind, insects and plant parts used for propagation.

## Favourable Conditions

- Temperature- 21 to 23$^0$C
- Warm cloudy and rainy weather

## Integrated Management

- Grow resistant varieties.
- Avoid wounding the crown and control root cheming insects.
- Foliar spray with Bordeaux mixture (0.1%), Streptocylin or Terramycin.
- Regulate soil fertility to reduce succulence of new growth.
- Deep seeds or root systems of nursery seedlings or root stocks upto crown region in *A. tumefaciens* strain K84 suspension.

## MODEL QUESTION PAPER

### Section 'A'

### Long answer questions

1. Explain in detail the occurrence, importance, symptoms, etiology, disease cycle and management of scab of apple.
2. Briefly describe the diagnostic symptoms and integrated management of any two of the following apple diseases :

   a. (i) Powdery mildew (ii Fire blight

   (iii) Crown gall
3. 3. Describe the powdery mildew of apple in the following headings:

   (i) Pathogen (ii) Symptoms

   (iii) Disease cycle (iv) Disease management
4. Describe in detail the most distinguishing symptoms, causal organism, disease cycle and integrated management of crown gall of apple.
5. Give the diagrammatic representation of the disease cycle of the following diseases of apple

   (i) Powdery mildew (ii) Scab

   (iii) Fire blight
6. Name two bacterial diseases of apple with their causal organism. Describe etiology, disease cycle and management of any one disease

## Section 'B'

### Short answer questions

1. Write down the etiology of powdery mildew of apple.
2. Which fungal pathogen causes crown gall of apple? Give its systematic position.
3. Write down the integrated management practices of scab of apple.
4. Describe the causal organisms and mode of penetration of firelight of apple.
5. Why the downy mildew disease of apple is historically important ?
6. Suggest some chemicals for management of scab of apple.
7. Which type of disease can be effectively managed by mixed cropping.

## Section 'C'

### Objective type questions

Fill in the blanks with suitable word (s)

1. Apple scab is caused by ________________ is telomorph, whereas, anamorph of the fungus is ________________.
2. Apple scab disease perinnates in ______________and primary infection is caused by ________________.
3. Leaf ____________period and ______________plays important role in primary infection of apple scab.
4. Secondary source of inoculum for scab development is ___________.
5. Application of__________to leaf litter during fall (when more than 80% leaves have fallen) hasten decomposition of leaves due to microbial action and reduces primary inoculum.
6. In case of fire blight of apple, tips of young infected shoots wilt, forming a very typical __________________symptom.
7. The fire blight bacteria ______________in bark tissues along the edges of cankers caused by infections during previous years.
8. The *Erwinina amylovora* penetrate the tree through ___________or ________________.
9. Secondary infection of *Erwinina amylovora* arises from ________of fresh infections.
10. *Podosphaera leucotricha* overwinters in the ______________that had been infected in the preceding growing seas

## Section 'D'

## State whether the following statements are True or False

1. *Venturia inaequalis* ia an obligate parasite.
2. *Erwinia amylovora* is a Gram +ve bacteria.
3. The fungus *Venturia inaequalis* is bipolar heterothallic.
4. The ascocarp of pathogen *Oidium farinosum* is known as cleistotheica.
5. White floury patches symptom appear on both side of the leaves of apple in powdery mildew disease.
6. *Erwinia amylovora* pv.*pyri* has the distinction of being the first bacterium of plants shown to be a pathogen.
7. Plasmid carries the genes for gall or tumour induction and is called tumour inducing plasmid (Ti-Plasmid).
8. *Agrobacterium tumefaciens* as an endospore forming bacteria.

## Answer

1. *Venturia inaequalis* (telomorph), *Spilocaea pomi* (Anamorph)
2. Pseudothecia, Ascospores
3. wetness, temperature
4. airborne conidia
5. 5% urea
6. shepherd's crook
7. Overwinter
8. natural openings or wounds
9. Ooze
10. dormant buds

**Fig. 1:** Scab

**Fig. 2:** Powdery mildew

**Fig. 3:** Crown gall

**Fig. 4:** Fire blight

**Plate 12:** Photograph showing symptoms of major diseases of apple *(See colour version on page 209)*

# 13

# Diseases of Peach and Their Management

| Sl No | Name of Diseases | Causal Organism |
|---|---|---|
| 1 | Leaf curl | *Taphrina deformans* (Berk) Tulansne |
| 2 | Leaf spot | *Phyllosticta cerasicola* Speg. |
| 3 | Powdery mildew | *Sphaerotheca pannosa* (Wallr.) Lév, *Podosphaera leucotricha* (Ellis &Everh.) E.S. Salmon |
| 4 | Whiskar rot | *Rhizopus stolonifer* Ehrenb |

## Leaf Curl

### *Taphrina deformans*

### Disease Symptoms

- Initially yellowish to reddish areas appear on leaf blade of young leaves in the spring.
- These areas progressively thicken and pucker along the midrib causing the leaf to curl.
- Infected leaves often develop red to purple color. Later on white to silvery coating of spores develop on the upper leaf surface and the affected leaves become crisp and brittle.
- Finally, infected leaves abscise prematurely or may sometimes remain attached , gradually turning dark brown on severly infected trees.
- Fruit infection is characterized by irregular, raised, wrinkled and reddish lesions.
- Infected flowers and fruits drop rapidally (Fig-1).

### Etiology

### Scientific Classification

Kingdom : Fungi

Division : Ascomycota

Class : Archiariomycetes

Order : Taphrinales

Family : Protomycetaceae

Genus : *Taphrina*

Species : *deformans*

- Mycelium: septate with typically binucleate cells, intercellular, mostly subcuticular,sub epidermal or may proceed deeper into the tissues.After maturity, the hyphal cells break up , become free form the neighbouring ones and are called chlaymydospores or ascogenous cells.
- Asci : naked clynideric, clavate, rounded or truncate at the apex and have a stalk cell size, measuring 17- 56x5-7μm and form compact and white hymenium.
- Ascus: contains eight ascopores which are round, ovate, elliptic and measure 3-7 μm in diameter.
- Ascospores: multiply by budding inside to outside the ascus, producing condia.
- Blastospores: (bud conidia) formed by budding from ascospores are oval, minute and measure 2.5 x4.8 μm, may germinate to produce mycelium.

## Disease Cycle

- Pathogen survives as blastospores in the crevices of the brak and unde the bud scales of the tree.
- Priamry infection occur in spring so called as 'spring time' when the bud swell and the first leaf emerges from it.
- Blastospores germinate and infect the yound leaves.
- Hyphae that grow intercellularly stimulate the adjacent cells to form the the neoplastic growth, rsulting in leaf curl symptoms.
- The hyphae later aggregate below the cuticle and breakup to form ascogenous cells, which develops into an ascus.
- Ascospores are released forcibly, multiply by budding inside and outside the ascus forming blastospores.
- These collect on the leaf surface giving rise it a powdery and wrinkled appearance.
- There is no secondary infection

## Favourable conditions

- Relative humidity- >85%
- Optimum Temperature – 15- $20^0$ C

## Integrated Management

- Select resistant varieties like Stark Early Giant, Starking Felicious, Tasis Semisto, World Earliest, Shan-e-Punjab, Bed Wills early, July Alberta when ever possible.
- Foliar spray with sulfur @ 0.2% or Mancozeb @ 0.2%,Blue copper or Blitox 50 @ 0.3percent . For best results, trees should be sprayed to the point of runoff or until they start dripping.
- Keep the ground beneath the trees raked up and clean, especially during winter months.
- Prune and destroy infected plant parts as they appear.
- If disease problems are severe, maintain tree health and vigor by cutting back more fruit than normal, watering regularly (avoiding wetting the leaves if possible) and apply an organic fertilizers high in nitrogen.

## MODEL QUESTION PAPER

### Section 'A'

### Long answer questions

1. Explain in detail the occurrence, importance, symptoms, etiology, disease cycle and management of leaf curl.
2. Briefly describe the diagnostic symptoms and integrated management of any two of the following peach diseases :

   (i) Powdery mildew (ii) Leaf curl

   (iii) Leaf spot
3. Describe the powdery mildew of peach in the following headings:

   (i) Pathogen (ii) Symptoms

   (ii) Disease cycle (iv) Disease management\

### Section 'B'

### Short answer questions

1. How can be manage peach leaf curl.
2. What are leaf curl disease in peach and how do we manage it?
3. What are some of the natural ways of managing powdery mildew?
4. How can we manage powdery mildew disease in peach?
5. What are leaf spot disease in peach and what are their symptoms?
6. How does one manage leaf spot?

## Section 'C'

### Objective type questions

Fill in the blanks with suitable word (s)

1. Powdery mildew of peach is caused by ……………………….. ..
2. Leaf curl of peach is caused by ……………
3. Foliar spray with sulfur fungicide control --------------- disease of peach.
4. The fungus *Taphrina deformans* belongs to family-----------------
5. *Taphrina deformans* survives as ---------------- on various parts of the peach.

## Section 'D'

### State whether the following statements are True or False.

1. *Taphrina deformans* survives as blastospores on various parts of the peach.
2. *Taphrina deformans* belongs to class Archiarcomycetes.
3. The ascus of *Taphrina deformans* contains eight ascopores.
4. There is no secondary infection in leaf curl disease.

**Fig. 1:** Leaf curl

**Plate 13:** Photograph showing symptoms of major disease of peach *(See colour version on page 209)*

# 14

# Diseases of Strawberry and Their Management

| Sl No | Name of Diseases | Causal Organism |
|---|---|---|
| 1 | Leaf spot/Common leaf spot | *Mycosphaerella fragariae*(Tul.) Lindau |
| 2 | Powdery mildew | *Podosphaera aphanis* |
| 3 | Vertillium wilt | Verticillium spp. |
| 4 | Botrytis blight | *Botrytis cinerea* |

## Leaf Spot

*(Mycosphaerella fragariae teleomorph, Ramularia brunnea Peck anamorph).*

### Disease Symptoms

- Older leaves usually show purple spots (3-6 mm in diameter) on the upper surface, sometimes with a slightly darker halo.
- As the spots mature, they become white to gray and surrounded by a brownish halo.
- A typical lesions, uniformly brown without darker border or lighter centers, may form on young leaves.
- Later on, the whole leaf is covered by numerous lesions and become chlorotic, withered and die.
- Fruits are generally not directly infected, but the loss of foliage can reduce their quality and yield (Fig-1).

### Etiology

### Scientific Classification

Kingdom : Fungi

Division : Ascomycota

Class : Dothideomycetes

Order : Capnodiales

Family : Mycospharellaceae

Genus : *Mycosphaerella*

Species : *fragariae*

- Teleomorph *Mycosphaerella fragariae* reproduces by forming perithecia that are black, partially embedded, globose (100-150 µm) and minutely ostiolate.
- Asci: Size 30-40 x 10-15 µm, cylindrical to clavate, have short stalks, and contain 8 spores, are formed in small clusters.
- Ascospores: Size12-14 x 3-4 µm, hyaline and two-celled with a median septum with each cell generally containing two oil drops.
- Anamorph of *M. fragariae, Ramularia brunnea* produces conidia that are elliptical to cylindrical (20-40 x 3-5 µm) hyaline, and zero- to four-septate.
- Conidia are often formed in short chains and are borne on short, hyaline, unbranched, and frequently curved conidiophores.
- Hyphae of *M. fragariae* may aggregate and form sclerotia which resemble perithecia in size and shape but lack a cavity.

## Disease cycle

- Pathogen overwinter on infected leaf debris on the soil.
- Resting spores produces conidia which are the source of primary infection.
- Secondary spread occurs through spores that are carried to new leaves by rain splashes and wing.

## Favourable conditions

- Relative humidity- >85%.
- Optimum Temperature – Around $25^0$ C.
- Prolonged leaf wetness.

## Integrated Management

- Use seeds for planting from certified sources.
- Plant resistant varieties, if available in area.
- Plant in well drained light soil with good air circulation.
- Clear the fields and surroundings of weeds.
- Apply a balanced fertilizer program without excess nitrogen.
- Foliar spray with Chlorothalonil 75% WP @ 0.2% or Copper oxychloride

50% WP @ 0.3% or Mancozeb 75% WP@ 0.2% or Hexaconazole 2% SC @ 0.1% as required depending upon crop stage (second spray after 15 days interval) .

- Do not water in the evening to avoid high humidity conditions.
- Remove infected plants and crop debris and burn or bury them at some distances of the field.
- Do not work in field when foliage is wet and avoid injuries the plants.

## MODEL QUESTION PAPER

### Section 'A'

### Long answer questions

1. Describe the leaf spot of Strawberry in the following headings;

   (i) Pathogen (ii) Symptoms

   (ii) Disease cycle (iv) Disease management.

2. Describe in detail the most distinguishing symptoms, causal organism, disease cycle and integrated management of leaf spot of strawberry.

3. Give the diagrammatic representation of the disease cycle of the following diseases of grapes

   (i) Leaf spot/Common leaf spot (ii) Powdery mildew

   (iii) Vertillium wilt

### Section 'B'

### Short answer questions

1) Write down the etiology of leaf spot of strawberry .
2) Which fungal pathogen causes leaf spot of strawberry? Give its systematic position.
3) Write down the integrated management practices of leaf spot of strawberry .
4) Suggest some chemicals for management of leaf spot of Strawberry .

### Section 'C'

### Objective type questions

ii. Fill in the blanks with suitable word (s)

1) Leaf spot of strawberry is caused by ......................... ..
2) The anamorphic stage of *Mycosphaerella fragariae* is ......

3) Teleomorph *Mycosphaerella fragariae* reproduces by forming ---------------that are black, partially embedded, globose and minutely ostiolate.
4) Hyphae of *M. fragariae* may aggregate and form ------------------- which resemble perithecia in size and shape but lack a cavity.
5) The fungus *Gloesporium ampelophagum* belongs to family------------------

**Fig. 1:** Leaf Spot

**Plate 14:** Photograph showing symptoms of major disease of Strawberry
*(See colour version on page 209)*

# 15

# Diseases of Potato and Their Management

| Sl.No | Name of Disease | Causal organism |
|---|---|---|
| 1 | Late blight | *Phytophthora infestans* (Mont.) de Bary |
| 2 | Early blight | *Alternaria solani* Sorauer |
| 3 | Black scurf | *Rhizoctonia solani* J.G. Kühn |
| 4 | Common scab | *Streptomyces scabies* Lambert and Loria |
| 5 | Leaf roll | Potato leaf roll virus |
| 6 | Mosaic | Potato virus X &Potato virus Y |

## Late Blight

### *Phytophthora infestans*

- Phytophthora in Greek means "plant destroyer"(phyto=plant, phthora= destroyer).
- Late blight is a most destructive disease of potato and is also of great historical importance.
- It was responsible for the Irish Potato Famine of 1843-50, the birth of plant pathology.
- It also infects tomato.
- India, the disease becomes serious occasionally on hills. However it occurs frequently in the plains (Figure-1).

## Diagnostic Symptoms

- Disease damages leaves, stems and tubers.

## Symtom on leaves

- Dark brown spots appear on leaf tips and margins.Inhumid conditions, these spots become water soaked leasions.
- Spots turn into transparent wounds.
- White fungus cover the underside of leaves
- Leaves wilt and die off.

### Symptom on stem

- Affected stems begin to blacken from their tips, and eventually dry out.
- Severe infections cause all foliage to rot, dry out and fall to the ground, stems to dry out and plants to die.

### Symptom on tuber

- Affected tubers display dry brown-colored spots on their skins and flesh.
- This disease acts very quickly. If it is not controlled, infected plants will die within two or three days.

### Etiology

### Scientific Classification

Kingdom : Fungi

Division : Eumycota

Class : Oomycetes

Order : Peronosporales

Family : Pythiaceae

Genus : *Phytophthora*

Species : *infestans*

- Mycelium: endophytic, coenocytic and hyaline which are inter cellular with double club shaped haustoria type.
- Sporangiophores: hyaline, branched intermediate and thick walled.
- Sporangia: thin walled, hyaline, oval or pear shaped with a definite papilla at the apex. Sporangium may act as a conidium and germinate directly to form a germ tube.
- Zoospores: biflagellate possess fine hairs while the other does not.
- Sexual reproduction in *Phytophthora* is not very common.

### Disease cycle

- Pathogen survives in soil, infected tubers and in plant debris.
- Infected tubers of the previous season lying in the field or from storage, serve as the spurce of primary inoculums.
- Under cold and humid conditions, sproangiophores and sporangia are formed on the infected tubers.

- The sporangia , which are lemon shaped with a beak, are dispersed by wind or soil water.
- They germinate directly by forming a germ tube, like a conidium, or indirectly by forming zoospores, depending on temperature and availability of water.
- Sporangia germinate directly at warm temperatures (20$^0$C or more) and when water is scare. Zoospres are formed at lower temperature (about 12$^0$C) and in presence of enough water.
- Germ tube formed by the sporangium grows and forms an appresorium at the tip.It forms an infection hypha that is pushed into the leaf by force generated by the turgor pressure in the appresorium.
- In indirect germination, the laterally biflagellate, kidney shaped zoospores that are released by the sporangium, settle down on the infection site, withdraw the flagella, and form a cyst.
- A germ tube emerges and infects the leaf or tuber.
- Pathogen colonizes the intercellular spaces and grows on the nutrients from the dead cells.
- Zoosporangia cause secondary infections, directly or through zoospores till plants are available.

## Favourable conditions

- High relative humidity (>90% ) coupled with suitable temperature.
- Optimum temperature for germination of the sporangia by zoospores as 12-13$^0$C and by germ tube 24$^0$C.
- Sporulation can occurs at any temperature from 9-26$^0$C.
- Optimum temperature for growth of mycelium is 16-18$^0$C.

### *Dutch rules*

- Night temperature below the dew point for 4 hours or more.
- Night temperature not below 10°C.
- Cloudiness on the next day.
- Rainfall at least 0.1mm on the following day.

## Integrated Management

- Soil solarization during summer.
- Field sanitation, rogueing.
- Avoid water logged conditions in the field.
- Follow crop rotation.

- Apply manures and fertilizers as per soil test recommendations.
- Start to grow ecological engineering plants.
- Sow/plant 4 rows of maize, sorghum, bajra (pearl millet) around the potato crop field as a guard/barrier crop.
- Select healthy tubers for planting.
- Delayed harvesting.
- High ridging to about 10-15cm height reduces tuber infection.
- Apply potasic fertilizer.
- Grow resistant varieties such as Kufri Jyothi, Kufri Badshah, Kufri Jeevan, Kufri Sherpa, etc.
- Dip sprouted tubers in 0.25% Mancozeb or 0.2% Metalaxyl or treat tuber with Carbendazim 25% + Mancozeb 50% WS @ (1.5 + 3.0) to (1.75 + 3.5) for 10 Kg seed (tuber) for 30 minute.
- Folair spraywith Chlorothalonil 75% WP @ 0.2% or Copper oxychloride 50% WP @ 0.3% or Mancozeb 75% WG @ 0.2% or Propineb 0.3% or Cymoxanil 8% + Mancozeb 64% WP @ 0.2% or Famoxadone 16.6% + Cymoxanil 22.1% SC @0.1% or Metalaxyl M 4% + Mancozeb 64% WP @ 0.25% or Metalaxyl 8% + Mancozeb 64% WP @ 0. 25% or Metiram 55% + Pyraclostrobin 5% WG @0.3% or Azoxystrobin 23% SC@ 0.1%, if required repeat after 10-15 days.

## Early blight

***Alternaria solani***

## Diagnostic Symptoms

- The disease is called "early" blight because it occurs earlier than the "late" blight, which occurs at the flwering stage.
- Cirucular or angular pale brown spots appear on the leaves.
- These are circular when formed in between the veins, and angular when bordered by veins.The spots subsequently get covered with grayish fungal growth.
- The tissue turn necrotic and concentric rings of raised and depressed necrotic zones are formed, giving the typical "target board" appearance. This is the most characteristic symptom of early blight.
- In dry weather , the leaves show curling where as in wet weather, affected areas coalesce to form big, rotting patches, which get shriveled.
- Tuber infections are evident after several months of storage, as dark sunken areas with raised dark brown margins appear on the tuber(Figure-2).

## Pathogen

## Scientific Classification

Kingdom : Fungi

Phylum : Ascomycota

Class : Dothideomycetes

Subclass : Pleosporomycetidae

Order : Pleosporales

Family : Pleosporaceae

Genus : *Alternaria*

Species : *solani*

- Mycelium: light brown or olivaceous which become dark coloured with age.
- Hyphae: branched, septate and inter and intra cellular.
- Conidiophore: emerge through the stomata or between the epidermal cells, broader than the hyphae.
- Conidia: club shaped with a long beak which is often half the long of the whole conidium. Lower part brown while the neck is colorless,body divided by 5 – 10 transverse septa and there may or may not be a few longitudinal septa.

## Disease cycle

- Pathogen overwinters in infected plant debris, infected tubers or in the soil.
- Conidia or the new conidia found on the overwintered mycelium bring about the primary infection of the succeeding potato crop.
- Secondary infection occurs through conidia formed on the spots developed due to primary infection, are transported by water, wind, insects, other animals including man, and machinery.
- Infection occurs through the stomata but direct penetration may also take place.

## Favourable condition

- Dry warm weather with intermittent rain .
- Poor vigor.

- Temperature: 25-30°C.
- Poorly manured crop.

## IntegratedManagement

- Disease free seed tubers should be used for planting.
- Plant on raised beds to improve drainage.
- Apply balanced fertilizer.
- Removal and destruction of infected plant debris.
- Folair spray Chlorothalonil 75% WP @ 0.2% (second spray after 14 days interval) or Copper oxychloride 50% WP @ 0.3% or Mancozeb 75% WP@ 0.2% or Hexaconazole 2% SC @ 0.1% (second spray after 21 days interval) or Kitazin 48% EC @ 0.20% as required depending upon crop stage and plant protection equipment used (second spray after 15 days interval) .

## Black Scurf

***Rhizoctonia solani (Perfect stage-Thanetophorus cucumeris)***

## Diagnostic Symptoms

- Raised black spots, irregular in size or shape, appear on the surface of the potato tubers (scurfs).
- These black marks can be rubbed or scraped off easily.
- Brown sunken patches with white fungal growth on upper roots and new shoots.
- Withering and discoloration of leaves.
- If the rot girdles the stem and blocks water and nutrient transport, the leaves becomes discolored and withered(Figure-3).

## Pathogen

## Scientific Classification

Kingdom : Fungi

Division : Eumycota

Sub- division : Deuteromycotina

Class : Hyphomycetes

Order : Agonomycetales

Family : Agonomycetaceae

Genus : *Rhizoctoniza*

Species : *solani*

- Mycelium : hyalinc when young and brown at maturity.
- Hyphae: septate and branched with a characteristic constriction at their diagnostic junction with the main hyphae.
- Sclerotia are black.
- Basidium: four sterigmata each with a basidiospore at the end.
- Basidiospore:s are hyaline, elliptical to obovate and thin walled. They are capable of forming secondary basidiospores.

**Disease Cycle**

- Pathogen is soil and seed borne, remain in soil and plant debris including infected tubers.
- Sclerotia on the seed tubers is the primary source of infection of the subsequent crop raised with these tubers.
- On return of favourable conditions the mycelium present in the soil may develop producing new hypae.The new hypha infects fresh tubers, germinating buds and stem.

**Favourable Condition**

- Moderately cool, wet weather
- Optimum temp for infection is 18 °C.
- Sandy soil .

**Integrated Management**

- Use disease free seed tubers.
- Avoid planting early in the season.
- Prepare shallower furrows to allow for earlier shoot emergence from the soil.
- Practice crop rotation.
- Water plant adequately , especially during dry periods.
- Soak seed tubers in Carbendazim + Mancozeb (0.25%) one day before sowing. The tubers should be thoroughly rinsed and dried in shade.
- Well sporulated tubers planted shallow.
- Disease severity is reduced in the land is left fallow for 2 years.

## 4. Mosaic

Mosaics of potato include a large number of distinct viral diseases.The overlapping of symptoms is invoked by different viruses individual or combination such as super mild, typical green or yellow mosaic and mild (Figure-4). and severe mosaic

| Name of diseases | Mild mosaic | Severe mosaic | Regose mosaic |
|---|---|---|---|
| Causal Organism | Potato virus X | Potato virus Y | Potato virus X &Potato virus Y |
| Diagnostic symtoms | • Light green mosaic pattern on leaves<br>• Mottled areas with small, brown dots<br>• Necrotic leaf tips, stunting, and death of plants | • Yellow to dark green mosaic pattern on leaves-starting from the tip<br>• Black necrotic spots and lines on leaves and shoots<br>• Stunted growth | • Folliage mottled, severly wrinkled, puckered, and reduced in size.<br>• Whole plant becomes dwarf.<br>• Leaflet margins roll downwards<br>• Rugosity, abnormal hariyness of the leaves and dwarfing of the plants |
| Transmitted by | • Direct contact, contaminated tools or Grass hopper | • Aphids | • Aphids<br>• Infected tubers |

### Integrated Management

- Plant only certified seeds.
- Use tolerant/resistant varieties.
- Monitor the field and remove/destroy all infected plants.
- Donot plant potatoes near susceptible hosts.
- Avoid injuries to plants.
- Disinfect your tools.
- Spray NSKE 5%.
- Apply Thiamethoxam 25% WG @ 40 g in 200 l of water/acre or for aphid control.
- Foliar spray Filpronil 5.0%SC @ 0.4L / acre (202 liter of water).
- Foliar spray Dimethoate 30% EC @ 264 ml in 200- 400 l water/acre.

### c) Leaf roll

***Potato leaf roll virus(PLRV)***

### Diagnostic Symptoms

- Inward rolling of leaves which may progress upward.
- Rolled leaflets become still, thick and leathery.
- Internodes may be shortened.
- Stolen becomes shorter and the number and size of the tubers are reduced.
- Phloem necrosis of tubers in some varieties also reported.
- High infection levels reduce tuber yield and marketability (Figure-5).

### Causal Organism

### Classification

Viruses

Nidovirales

Lutoviridae

Polerovirus

- Potato leaf roll virus (Potato virus-1 or Solanum virus-14).

### Disease Cycle

- Infected tubers, volunteer and weed hosts serve as reservoir of primary inoculums.
- Plant raised form infectd tubers show disease symptoms.
- Virus remains in the phloem and circulates to other parts of the plant.
- Aphids; (*Myzus persicae*) feeding on phloem , transmit the virus to other plants in the fields, and thus cause secondary infections.
- Aphid is the most important vector and is responsible for 95% of the secondary infections.
- Virus is persistent i.e the vector transmits the virus through out its life.
- RLRV is the only virus infecting potato that is persistant in the vector.
- Secodary infection takes place when contaminated tubers are planted and potato plants grew from them.

## Favourable Condition

- Warm and dry weather.
- Maximum secondary infections occur during spring, when winged aphids are most prevalent.
- Aphids double their number in less than two days during warm and dry weather.

## Integrated Management

- Use disease free seed tubers for planting.
- Plant resistant/tolerant varieties if available.
- Monitor field , pick and destroy diseased plants.
- Destroy weeds and volunteer plants that can host the virus and feed the aphids.
- Apply Imidacloprid or Thiamethoxam 25% WG @ 40 g in 200 l of water/acre or soil drenching of Thiamethoxam 25% WG @ 80 g in 200 l water/ acre for aphid control.

## MODEL QUESTION PAPER

### Section 'A'

### Long answer questions

6. Explain in detail the occurrence, importance, symptoms, etiology, disease cycle and management of late blight of potato.
7. Explain in detail the diagnostic symptoms, etiology, disease cycle and management of early blight of potato.
8. Briefly describe the diagnostic symptoms and integrated management of any two of the following potato diseases :

   (i) Black scurf (ii) Common scab

   (iii) Late blight
3. Describe the black scurf of potato in the following headings:

   (i) Pathogen (ii) Symptoms

   (ii) Disease cycle (iv) Disease management
4. Describe in detail the most distinguishing symptoms, causal organism, disease cycle and integrated management of mosaic of potato .
5. Give the diagrammatic representation of the disease cycle of the following diseases of potato (i) Late blight of potato (ii) Early blight of potato.

6. Name two bacterial diseases of potato with their causal organism. Describe etiology, disease cycle and management of any one disease.
7. Name two fungal diseases of potato with their causal organism. Describe etiology, disease cycle and management of any one disease.

**Section 'B'**

**Short answer questions**

8. Write down the infection process in late blight of potato.
9. Describe the difference between late blight and early blight of potato.
10. Write down the integrated management practices of late blight of potato.
11. Why the late blight disease of potato is historically important ?
12. Suggest some new generation chemicals for management of late blight of potato.
13. Describe the disease cycle of leaf roll of potato .

**Section 'C'**

**Objective type questions**

Fill in the blanks with suitable word (s)

1. Black heart of potato caused by ------------------------ deficiency.
2. Due to excessive application of nitrogen fertilizer, an irregular cavity in the centre develop called ---------------------------
3. Late blight disease of potato is caused by _______________.
4. Early blight disease of potato is caused by ------------------------
5. Black scurf disease of potato is caused by---------------------------------
6. Excess Mg produces type of toxicity in in potato stem called------------
7. Potato leaf roll virus is transmitted by ----------------------------
8. Scab disease incidence condition reduces in -------------------------------
9. Irish famine came due to disease --------------------in Ireland.
10. Domestic quarantine is done mainly to protect from-----------------------
11. ------------------------ is a technique of multiplication of viral disease free seed in northern plain of the country.
12. The perfect stage of *Rhizoctonia solani is* ----------------------------------

## Section 'D'

**State whether the following statements are True or False.**

9. Kufri Kanchan and Kufri Jyoti is a wart resistant variety of potato.
10. Phytophthora in Greek means "plant destroyer.
11. K Thenamalai is cyst nematode and late blight resistant variety of potato.
12. Irish Potato Famine of 1845-50 leads to the birth of plant pathology.
13. Sexual reproduction in Phytophthora is not very common.
14. Sporangia of Phytophthora is thin walled, hyaline, oval or pear shaped with a definite papilla at the apex.
15. Dutch rules is related with late blight of potato.
16. Target board" appearance is the most characteristic symptom of early blight of potato.
17. The condia of *Alternaria solani* has both transverse septa and longitudinal septa.
18. Mild mosaic of potato is caused by Potato virus X.

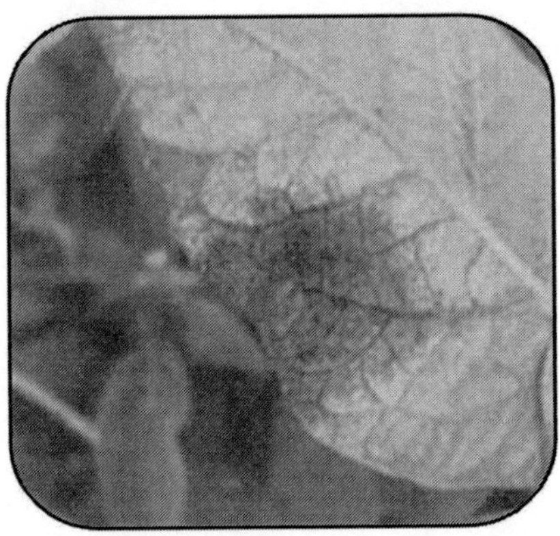

**Fig. 1:** Late blight

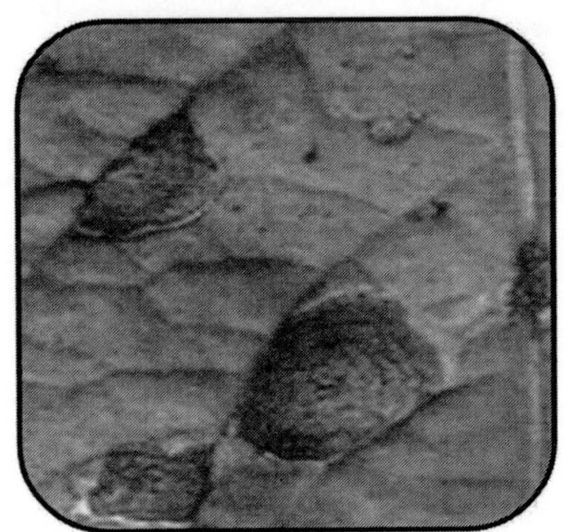

**Fig. 2:** Early blight

**Fig. 3:** Black scurf

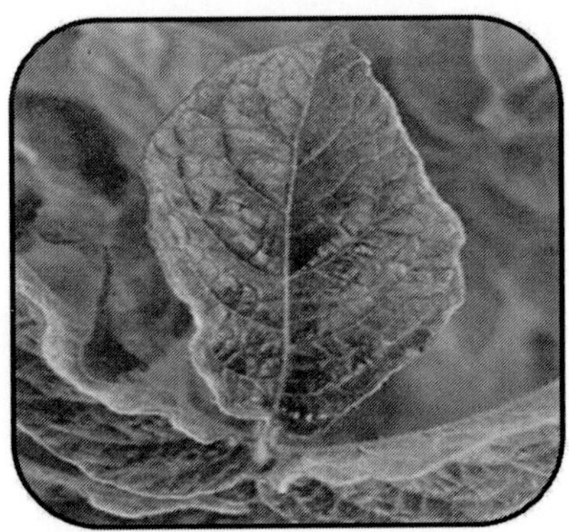

**Fig. 4:** Mosaic

**Fig. 5:** Leaf curl

**Plate 15:** Photograph showing symptoms of major disease of potato *(See colour version on page 210)*

# 16

# Diseases of Cucurbits and Their Management

| Sl.No | Name of Diseases | Causal organism |
|---|---|---|
| 1 | Downy Mildew | *Pseudoperonospora cubensis* (Berkeley & M. A. Curtis) |
| 2 | Powdery mildew | *Erysiphe cichoracearum* DC, *Sphaerotheca fuligenia* (Schltdl.), |
| 3 | Anthracnose | *Colletotrichum orbiculare* |
| 4 | Fusarium wilt | *Fusarium oxysporum* Schlecht |
| 6 | *Cercospora* leaf spot | *Cercospora citrullina, C. melonis, C. lagenarium* |
| 7 | Alternaria Blight | *Alternaria cucumerina,* |

## Downy mildew

### *Pseudoperonospora cubensis*

### Diagnostic Symptoms

- Yellow, angular spots restricted by veins resembling mosaic mottling appear on upper surface of leaves .
- Corresponding lower surface of these spots shows a purplish downy growth in moist weather.
- Spots turn necrotic with age.
- Diseased leaves become yellow and fall down .
- Diseased plants get stunted and die .
- Fruits produced may not mature and have a poor taste (Figure-2)

### Etiology

### Scientific Classification

Kingdom : Fungi

Division : Eumycota

Class : Oomycetes

Order : Peronosporales

Family : Peronosporaceae

Genus : *Pseudoperonospora*

Species : *cubensis*

- Obligate parasite.
- Mycelium: coenocytic and intercellular with small ovate or finger like haustoria.Mycelium develop one to five sporangiosphores arise through the stomata.
- Sproangiophores: 200-300 μm long and 5-9 μm broad, dichotomous branching at acute angle on which sporangia borne.
- Sporangia: grayish to olivaceous purple, ovoid to ellipsoidal, thin walled with a distal papilla at distal end, measure 21-39x 14-23 μm and germinate producing biflagellate zoospores.
- Zoospores: 10 – 13 micron meter, biflagellate.
- Oospores : not common, spherical, rarely avoid to ellipsoid, light yellow and smooth walled and measure 19-22μm in diameter.

## Disease Cycle

- Pathogen survives on the diseased plant debris or wild cucurbits and cause infection through roots.
- Primary infection occurs thorugh fungal growth and sporangia produced on infected plant parts.
- Secondary infection occurs thorugh sporangia germinates and produce zoospores.

## Favourable conditions

- Relative humidity > 90%
- High soil moisture
- Frequent rains

## Integrated Management

- Deep ploughing of fields during summer.
- Soil solarization: Cover the beds with polythene sheet of 45 gauge (0.45 mm) thickness for three weeks before sowing for soil solarization which will help in reducing the soil pathogen.
- Apply Trichoderma spp. @ 2.5 kg/acre along with FYM.
- Trellising cucumbers.
- Avoiding overhead irrigation or irrigating only in the late morning hours will limit the amount of time that leaves are wet.

- Control alternate weed hosts (wild cucumber, golden creeper and volunteer cucumbers) in neighbouring fence rows and field edges.
- Foliar spray of Mancozeb + Carbendazim @0.25% or Cymoxanil 8% + Mancozeb 64% WP @ 0.25%, Famoxadonc 16.6% + Cymoxanil 21.1% SC @0.1% if required repeat after 10-15 days.

## Powdery Mildew

### *Erysiphe cichoracearum (Sphaerotheca fuligena)*

### Host range

- Pumpkins, bottle gourd, coccinia, cucumber, ridge gourd.

### Diagnstic Symptoms

- Whitish or dirty grey, powdery growth on foliage, stems and young growing parts.
- Superficial growth ultimately covers the entire leaf area.
- Diseased areas turn brown and dry leading to premature defoliation and death.
- Fruits remain underdeveloped and are deformed (Figure1).

### Etiology

### Scientific Classification

Kingdom: Fungi

Division : Eumycota

Class : Leotiomycetes

Order : Erysiphales

Family : Erysiphaceae

Genus : *Erysiphe/Sphaerotheca*

Species : *chichoracearum/fuliginea*

- Bultler in 1918 reported the occurrence of both the organisms on cucurbits.
- Prior to 1958, *E. cichoracearum* was considered to be primary causal organism throughout the world but now *S. fuliginea* is regarded as the principal causal organism. It is also more common and more aggressive than the former.

- *E. cichoracearum* is present only in the initial stages of the disease.
- The conidia of the two fungi are very similar and difficult to be distinguished from each other, but the presence of fibrosan bodies in the conidia of *S. fulginea* enables a firm identification of the fungus.

### *Erysiphe chichoracearum*

- Mycelium: colorless, profusely branched, spetate mycelia, develops haustoria inside the host cells to obtain nutrients.
- Conidiophores: erect, stout, small and unbranched producing conidia at the apex.
- Conidia: produced in chain, ellipsoidal or barrel shaped, measure 63.8 x 31.9 micron meter, disseminated by air.
- Cleistothecia:dark, spherical with myceloid appendages which contain 10 – 15 asci.
- Ascus: contains two rarely three ascospores which are single celled, hyaline, measuring 20-30x12-18 mm.
- Cleistotheical stage is not very common.

### *Sphaerotheca fuliginea*

The major point where it differs from that of Erysiphe is the sexual stage.

- Produces cleistothecia each containing only one ascus, which is broadly elliptic to subglobose measuring 50-80x30-60μm.
- Each ascus contains eight ascospores and each ascospores is ellipsoid to nearly spherical.

### Disease cycle

- Pathogen overwinter or survives on infected plant debris in the soil as cleistotheica.
- Cliestothecia, though formed, are not significant in disease initiation.
- Pathogen has a wide host range and therefore, is always available on one or the other collateral host.
- Conidial stage of the pathogen release conidia for primary infection on the spring or summer sown cucurbits.
- Secondary infection is caused by wind borne conidia produced during primary infection.

### Favourable conditions

- Optimum temperature- 30$^{0}$C.
- Dry conditions.

### Integrated Management

- Plant resistant/ tolerant varieties.
- Increasing air movement inside the canopy.
- Monitor fields regularly to assess the incidence of a disease.
- Foliar spray with Carbendazim + Mancozeb (025%) or Thiophanate methyl 70% WP @ 0.1% or Calixin 0.1% or Fluopyrum + Tebconazole.
- Crop rotation with non susceptible crops.
- Apply balanced fertilizers.
- Remove plant residue after harvest.

### Anthracnose

***Colletotrichum orbiculare (= C. lagenarium)***

### Diagnostic Symptoms

- Lesions can form on seedlings, leaves, petioles, stems and fruits of susceptible cucurbits.
- Sunken, elongated stem cankers are most prominent on muskmelon, though leaf and fruit lesions also occur.
- Large lesions girdle the stems and cause the vines to wilt.
- Stem cankers are less obvious on cucumbers, but leaf lesions are very distinct
- Watermelon foliage affected by anthracnose appears scorched; sunken fruit lesions are easy to recognize (Figure-4)

### Etiology

### Scientific Classification

| | |
|---|---|
| Kingdom | : Fungi |
| Division | : Eumycota |
| Sub- division | : Deuteromycotina |
| Class | : Coelomycetes |
| Order | : Melanconiales |
| Family | : Melanconiaceae |
| Genus | : *Colletotrichum* |
| Species | : *lagenarium* |

- Telemorph *Glomerella lagenarium* is rarely found in nature.
- Pathogen produces black stromata bearing black setae and hyaline conidiophores on host surface.
- Conidia: are produced successively by budding at the tips of conidiophores accumulating in pinkish mass, one celled hyaline, oblong to ovate 4-6 by 13-19 mm.

## Disease Cycle

- Pathogen is seed borne, can also survive in infected plant debris, volunteer plants.
- Conidia are disseminated by wind, rain, implements.
- Spores are splashed from leaf to leaf and plant to plant, during irrigation or rain events.
- Several disease cycles can occur in a single growing season, resulting in defoliation of severely infected plants
- Visible symptoms appear up to 72 h after infection.

## Environmental conditions

- Humid rainy water is essential for infection.
- Spore germination is optimum at 22-27$^0$C
- Relative humidity-100% for 24h.

## Integrated Management

- Use certified, disease free seeds.
- Choose resistant varieties, if available .
- Deep ploughing of crop residue immediately after harvest.
- Seed treatment with Carbendazim 2g/kg of seed.
- Foliar spray with Mancozeb 2g or Carbendazim 1 g/litre of water.
- Crop rotation in which no cucurbits are grown for at least one year period.
- Practice good sanitation of the field by plowing under fruits and vines at the end of each season.

## Fusarium wilt

### *Fusarium oxysporum Schlecht*

### Diagnostic Symptoms

- Initial symptom of the disease is clearing of the veinlets and chlorosis of the leaves.
- Younger leaves may die in succession and the entire plant may wilt and die in a course of few days.
- Soon the petiole and the leaves droop and wilt.
- In young trailing plant, symptom consists of clearing of veinlet and dropping of petioles. In field, yellowing of the lower leaves first and affected leaflets wilt and die.
- Symptoms continue in subsequent leaves. At later stage, browning of vascular system occurs.
- Plants become stunted and die (Figure-3).

### Etiology

### Scientific Classification

Kingdom : Fungi

Division : Eumycota

Sub- division : Deuteromycotina

Class : Hyphomycetes

Order : Hyphomycetales (Moniliales)

Family : Dematiaceae

Genus : *Fusarium*

Species : *solani*

- Pathogen overwinter or survives on infected plant debris in the soil as cleistotheica, microconidia are formed singly, hyaline and cylindrical.
- Macro conidia are cylindrical to falcate.
- Chlamydospores are globose to oval and rough walled. discoloured, particularly in the lower stem and roots.

### Disease Cycle

- Pathogen survives on seeds or plant debris and other host plant parts.

- Primary spread through seed , soil, water, seedling, workers etc.
- Secondary spread through irrigation water, wind from infected plant debris.

## Favourable Conditions

- Relatively high soil moisture and soil temperature.

## Management

- Deep ploughing of fields during summer.
- Use pathogen free seeds.
- Soil solarization: Cover the beds with polythene sheet of 45 gauge (0.45 mm) thickness for three weeks before sowing for soil solarization which will help in reducing the soil borne disease.
- Apply Trichoderma spp. @ 2.5 kg/acre along with FYM.
- Remove and destroy the infected plants and plant debris.
- Avoid water stagnation and maintain proper drainage.

# MODEL QUESTION PAPER

## Section 'A'

## Long answer questions

1. Explain in detail the occurrence, importance, symptoms, etiology, disease cycle and management of downy mildew of bottle gourd.
2. Briefly describe the diagnostic symptoms and integrated management of any two of the following cucurbit diseases :

   (i) Powdery mildew (ii) Downy mildew ( iii) Anthracnose
3. Describe the powdery mildew of cucrbits in the following headings:

   (i) Pathogen (ii) Symptoms

   (ii) Disease cycle (iv) Disease management\
4. Describe in detail the most distinguishing symptoms, causal organism, disease cycle and integrated management of anthracnose of cucurbits.
5. Give the diagrammatic representation of the disease cycle of the following diseases of cucurbits

   (ii) Powdery mildew (ii) Downy mildew

   (iii) Anthracnose

## Section 'B'

### Short answer questions

1. Write down the etiology of powdery mildew of bottle gourd.
2. Which fungal pathogen causes downy mildew of muskmelons? Give its systematic position.
3. Write down the integrated management practices of anthracnose of water melon.
4. Describe the causal organisms and mode of penetration of powdery mildew of cucurbits.
5. Suggest some new generation chemicals for management of downy mildew of cucurbits.

## Section 'D'

### Objective type questions

(a) Fill in the blanks with suitable word (s)

1. Downy mildew of bottle gourd is caused by ......................... ..
2. Anthracnose of water melon is caused by ..............
3. The anamorphic stage of anthracnose of cucurbits is ......
4. Foliar spray with sulfur fungicide control --------------- disease of water melon.
5. The fungicide metalaxyl is used for management of ---------------- disease .

(b) State whether the following statements are True or False.

1. *Pseudoperonospora cubensis* ia an obligate parasite.
2. The occurrence of both *Erysiphe* and *Sphaerotheca* on cucurbits was reported by Butler.
3. The ascus of *Sphaerotheca fuliginea* contains eight ascospores.
4. *Sphaerotheca fuligena* is perfect stage of *Erysiphe cichoracearum.*
5. *Erysiphe chichoracearum* overwinter on infected plant debris in the soil as cleistothecia.

**Fig. 1:** Powdery mildew

**Fig. 2:** Downy mildew

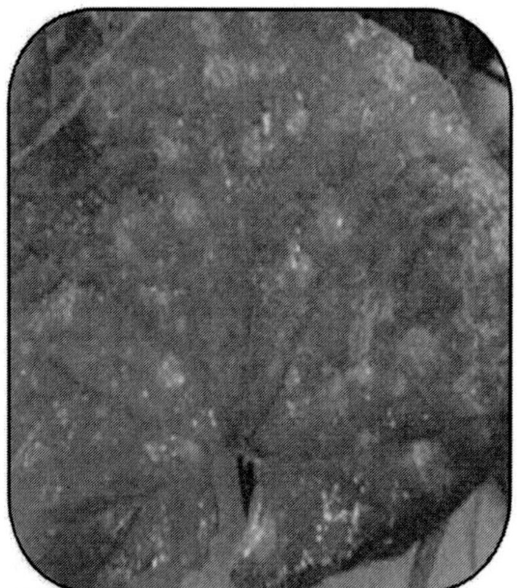

**Fig. 3:** Fusarium wilt

**Fig. 4:** Anthracnose

**Plate 16:** Photograph showing symptoms of major diseases of cucurbits *(See colour version on page 211)*

# 17

# Diseases of Onion and Garlic and Their Management

| Sl.No | Name of Disease | Causal organism |
|---|---|---|
| 1 | Purple blotch | *Alternaria porri* (Ellis) |
| 2 | Stemphyllium blight | *Stemphyllium cesicarium/vesicarium* Walla E.G.Simmons |
| 3 | Smut | *Urocystis cepulae* |
| 4 | Downy mildew | *Peronospora destructor* |
| 5 | Leaf blight (Blast) | *Botrytis* spp. |
| 6 | Black mould | *Aspergillus niger* |

## Purple blotch

### *Alternaria porri*

### Diagnostic Symptoms

- Disease symptoms occurs on leaf and flower stalks.
- Small, whitish, sunken lesions with purplish centre which rapidly enlarge and eventually girdle the leaf or seed stem.
- Usually the affected leaf or stem falls down and dies within 3 or 4 weeks.
- Infected plant fail to develop bulbs (Fig-1).

### Etiology

### Scientific Classification

Kingdom : Fungi

Phylum : Ascomycota

Class : Dothideomycetes

Subclass : Pleosporomycetidae

Order : Pleosporales

Family : Pleosporaceae

Genus : *Alternaria*

Species : *poori*

- Mycelium: is branched, coloured and septate.
- Conidiophores: arise singly or in groups, straight or flexuous, sometimes geniculate.
- Conidia: muriform, straight or curved, obclavate, tapering to a beak, golden brown in color and smooth, 8-12 transverse and several longitudinal septa, germinate by germ tube.
- Chlamydospres: survival structure

## Disease Cycle

- Pathogen survives in soil, infected bulbs and may persist in plant debris or on roots on weeds.
- Primary infection occurs by infected soil or inoculums present on plant debris.
- Secondary spread by conidia through rain or wind.

## Favourable conditions

- Warm humid weather with rains or heavy dew.
- Optimum Tmeperature-22$^0$C.
- Relative humidity-90%.

## Integrated Management

- Disease free bulb should be selected for planting.
- Use recommended dose of N and P fertilizers.
- Hot water soaking of seeds (50$^0$ C for 20 minutes).
- Seeds should be treated with Carboxin + Thiram @ 2.5 g/kg seed.
- Field should be well drained.
- Follow crop rotation with non host crops.
- Three foliar spray withDifenconazole 0.05% or Kitazin 0.1% or Copper oxychloride 0.25 %, if required repeat after 15 days.

## Stemphylium Blight

### *Stemphyllium cesicarium/vesicarium*

## Diagnostic Symptoms

- Small, water soaked, white to light yellow spots on leaves.
- Over time, sunken, elongated, brown blotches with tan to brown centers form.

- Large necrotic areas cause extensive blighting of the tissues.
- Concentric zones may also develop in their center.
- In advance stages, large necrotic areas form, which may girdle the leaf or seed stem,causing extensive blighting of the tissues.
- Disease on inflorescence stalk causes severe damage to seed crop(Fig-2).

## Etiology

### Scientific Classification

Kingdom : Fungi

Phylum : Ascomycota

Class : Dothideomycetes

Subclass : Pleosporomycetidae

Order : Pleosporales

Family : Pleosporaceae

Genus : *Stemphyllium*

Species : *vesicarium*

### Disease Cycle

- Pathogen survives in plant debris or soil.
- Priamry infection occurs by infected soil or inoculums present on plant debris.
- Secondary spread by conidia through rain or wind.

### Favorable Conditions

- Optimum temperature-18-25$^0$C
- Long period of leaf wetness(16 hours or more)

### Integrated Management

- Collect and burn the crop residues.
- Use certified seed.
- Select disease resistant cultivars.
- Crop rotation with non host crop
- Follow proper field drainage and reduce plant density.
- Ensure adequate field drainage before planting.
- Avoid excessive nitrogen applications.

- Foliar spray of Chlorothalonil @ 2 g/L or Azoxystrobin + Difenoconazole @ 1 ml/L or Boscalid + Pyraclostrobin@ 1 ml/L, if required repeat after 10 days.

## MODEL QUESTION PAPER

### Section 'A'

### Long answer questions

14. Explain in detail the occurrence importance, symptoms, etiology, disease cycle and management of purple, blotch of onion.
15. Briefly describe the diagnostic symptoms and integrated management of any two of the following onion diseases :

    (i) Purple blotch (ii) Stemphyllium blight
16. Describe in detail the most distinguishing symptoms, causal organism, disease cycle and integrated management of stemphyllium blight of onion
17. Give the diagrammatic representation of the disease cycle of the following diseases of onion

ii. Stemphyllium blight (ii) Purple blotch ( iii) Downy mildew

### Section 'B'

### Short answer questions

1. Write down the etiology of purple blotch of onion.
2. Which fungal pathogen causes stemphyllium blight of onion? Give its systematic position.
3. Write down the integrated management practices of purple blotch of onion
4. Suggest some new generation chemicals for management of downy mildew of cucurbits.

### Section 'C'

### Objective type questions

(a) Fill in the blanks with suitable word (s)

1. Downy mildew of onion is caused by ……………………….. ..
2. Purple blotch of onion is caused by …………..
3. *Alternaria poori* has ----------------------type of condia
4. *Stemphyllium cesicarium* belongs to family ---------------- '

5. The new generation fungicide commonly used for management of purple blotch is -----------------------------------

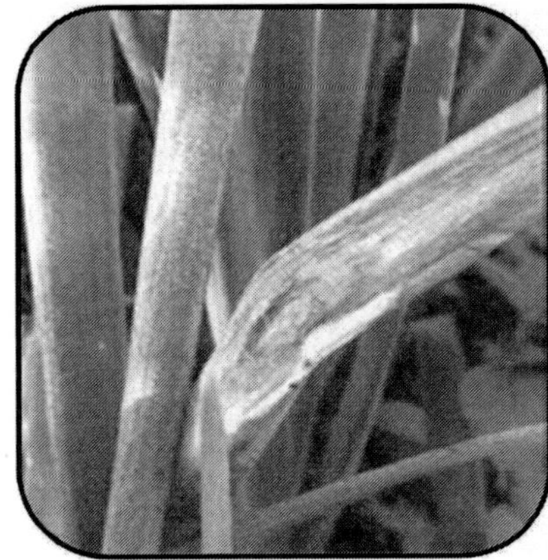

**Fig. 1:** Purple blotch **Fig. 2:** Stemphyllium blight

**Plate 17:** Photograph showing symptoms of major disease of onion *(See colour version on page 211)*

# 18

# Diseases of Chillies and Their Management

| Sl.no | Name of Diseases | Causal Organism |
|---|---|---|
| 1 | Damping off | *Pythium aphanidermatum* (Edson) Fitzp |
| 2 | Fruit Rot and Die Back | *Colletotrichum capsici* (Syd.) |
| 3 | Fusarium wilt | *Fusarium solani (Mart.) Sacc* |
| 4 | Powdery mildew | *Leveillula taurica* (Lév.) G. Arnaud |
| 5 | Viral diseases | |
| A | Mosaic | Chilli mosaic virus |
| B | Leaf curl | Tobacco leaf curl virus |

## 2. Anthracnose or Fruit Rot and Die Back

***Colletotrichum capsici***

### Diagnostic Symptoms

#### Dieback symptoms

- Small, circular to irregular, brownish black scattered spots appear on leaves .
- Severely infected leaves defoliate.
- Infection of growing tips leads to necrosis of branches from tip backwards.
- Necrotic tissues appear grayish white with black dot like acervuli in the center.
- Shedding of flowers due to the infection at pedicel and tips of branches.

#### Fruit symptoms

- Ripe fruits are more vulnerable to attack than green ones .
- Small, circular, yellowish to pinkish sunken spots appear on fruits.
- Spots increase along fruit length attaining elliptical shape.
- Severe infection result in the shrivelling and drying of fruits.
- Tissues around lesions will be bleached and turn white or greyish in colour and lose their pungency.

- On the surface of the lesions minute black dot like fruiting bodies called 'acervuli' develop in concentric rings and fruits appear straw coloured.
- Affected fruits may fall off subsequently. The seeds produced in severely infected fruits are discoloured and covered with mycelial mat (Figure-1).

## Etiology

### Scientific Classification

Kingdom : Fungi

Division : Eumycota

Sub- division : Deuteromycotina

Class : Coelomycetes

Order : Melanconiales

Family : Melanconiaceae

Genus : *Colletotrichum*

Species : *capsici*

- Mycelium: septate, colorless and inter and intra cellular.
- Hypha: collect beneath the epidermis and develop the fruiting bodies the acervuli.
- Acervuli: consists of setae, conidiophores and conidia.
- Setae: brown 1-5 septate, dark brown with light browing tips, rigid, hardly swollen at the base, slightly tapered towards the paler acute apex.
- Conidiophores: club shaped, unicellular, unbranched small, hyaline and densely arranged in fruiting body, produce conidia at their tips.
- Conidia: hyaline, unicellular, curved, with narrow ends and produced singly at the tips of the conidiophores.Conidia *en masse* appear pinkish.

## Disease Cycle

- Pathogen survives through mycelium in the plant debris which include infected twigs, stem and fruits .
- Perennating mycelium germinates in soil producing conidia, disseminated and cause primary infection on growing crop.
- Secondary spread take place through the wind borne conidia produced from primary infection.
- Conidia germinate within four hours in water and an appresoria is developed for further growth of the pathogen between the plant tissues.

## Favourable Conditions

- Optimum temperature -28$^0$ C.
- Relative Humidity- > 97%.
- Humid weather with rainfall at frequent intervals.

## Integrated Management

- Use healthy pathogen-free chilli seed.
- Removal and destruction of Solanaceous weed hosts and infected plant debris.
- Resistant varieties: Hisar Shakti, Hisar Vijay, TNAU Chilli Hybrid Co 1,Perennial, Bengal Green, S20-1, H-4, H-6, Lorai, etc.
- Seed treatment with Carboxin + Thiram 2.5g/kg.
- Early removal of affected plants.
- Transplants should be kept clean by controlling weeds and solanaceous volunteers in the vicinity of the transplant houses.
- Stagnation of water should not be allowed in nursery beds and fields .
- Field should have good drainage and be free from infected plant debris.
- Foliar spray with Copper oxy chloride 50% WP @ 0.3% or Difenoconazole 25% EC @ 0.025% or Tebuconazole 25% WG @ 0.2% or Azoxystrobin 23% SC @ 0.1% or Chlorothalonil @ 75% WP 0.1% if required repeat after 15 days.

## Fusarium wilt

## *Fusarium solani* (Mart.) Sacc

## Diagnostic Symptoms

- Initial symptom of the disease is clearing of the veinlets and chlorosis of the leaves.
- Younger leaves may die in succession and the entire may wilt and die in a course of few days. Soon the petiole and the leaves droop and wilt.
- In young plants, symptom consists of clearing of veinlet and dropping of petioles. In field, yellowing of the lower leaves first and affected leaflets wilt and die.
- Symptoms continue in subsequent leaves. At later stage, browning of vascular system occurs. Plants become stunted and die (Figure-2). .

## Etiology

### Scientific Classification

Kingdom : Fungi

Division : Eumycota

Sub- division : Deuteromycotina

Class : Hyphomycetes

Order : Moniliales

Family : Dematiaceae

Genus : *Fusarium*

Species : *solani*

- Mycelium is grayish white.
- Microconidia are formed singly, hyaline and cylindrical.
- Macro conidia are cylindrical to falcate.
- Chlamydospores are globose to oval and rough walled. discoloured, particularly in the lower stem and roots.

### Disease Cycle

- Pathogen survives on seeds or plant debris and other host plant parts.
- Primary spread through seed , soil, water, seedling, workers etc.
- Secondary spread through irrigation water, wind from infected plant debris.

### Favourable Conditions

- Optimum soil temperature-20-30$^0$C.
- Hot and dry summers followed by rains.

### Integrated Management

- Use pathogen free seeds.
- Remove and destroy the infected plants and plant debris.
- Adopt crop rotation.
- Avoid water stagnation and maintain proper drainage.
- Use of wilt resistant varieties like Phule Jyoti, Phule Mukta.
- Drenching with 1% Bordeaux mixture Captan + Hexaconazole @ 0.2% may give protection.·

- Seed treatment with 4g *Trichoderma viride* formulation or Carboxin + Thiram @ 2.5g/kg seed.
- Mix 2kg *T.viride* formulation with 50kg FYM, sprinkle water and cover with a thin polythene sheet. When mycelia growth is visible on the heap after 15 days, apply the mixture in rows of chilli in an area of one acre.

## Viral diseases

### Leaf curl

### Diagnostic Symptoms

- Leaves curl towards midrib and become deformed.
- Stunted plant growth due to shortened internodes and leaves greatly reduced in size.
- Flower buds abscise before attaining full size and anthers do not contain pollen grains.
- Virus is generally transmitted by whitefly. So control measures of whitefly in this regard would be helpful(Figure-3).

### Etiology

- Tobacco leaf curl virus causes the leaf curl disease.
- Virus is not sap transmitted and is also not seed borne.
- In nature, it is transmitted by the white fly (*Bemesia tabaci*)

### Integrated Management

- Selection of healthy and disease - free seed.
- Use resistant /tolerant varieties like Pusa Sadabahar, Arka Harita, Arka Meghana, ArkaSweta, Hisar Shakti, Hisar Vijay, Pant C-1
- Nursery beds should be covered with nylon net or straw to protect the seedlings from viral infection.
- Raise 2-3 rows of maize or sorghum as border crop to restrict the spread of aphid vectors.
- Apply Carbofuran 3G @ 4-5 Kg/acre in the mainfield to control sucking complex and insect vectors selectively.
- Seed treatment with Imidacloprid 70% WS @ 10 g/kg of seed.
- Apply Fipronil 5% SC @ 320-400 ml in 200 l of water/acre.
- Collect and destroy infected virus plants as soon as they are noticed.

## MODEL QUESTION PAPER

### Section 'A'

### Long answer questions

1. Describe diagnostic symptoms, etiology , disease cycle and management of damping off of chilli.
2. What are the two mechanisms proposed to explain pathologiocal wilting of chilli? Describe the plugging theory.
3. Name four diseases of chilli and their causal organisms?. Describe any two of them; one fungal and one bacterial.
4. Mention the fungal disease of chilli and their causal agents. Describe Fruit rot and Die back.
5. Explain in detail the occurrence, importance, symptoms, etiology, disease cycle and management of fruit rot and die back

### Section 'B'

### Short answer questions

1. How will you differentiate fruit rot and die back.
2. Write down the etiology of powdery mildew of chilli.
3. Write down the integrated management practices of fruit rot and die back of chilli.
4. Distinguish between pre emergent and post emergent damping off diseases.
5. Distinguish between soft rot and dry rot.

### Section 'C'

### Objective type questions

(a) **Fill in the blanks with suitable word (s)**

1. Leaf curl of chill is transmitted by ----------------
2. Anthracnose of chilli is caused by ..............
3. All wilt causing Fusaria placed together under the species ---------
4. The Fusarium macroconidia have a distinct ------------------- bearing some kind of heel is the distinctive feature of Fusarium.
5. Foliar spray with sulphur fungicide control --------------- disease of chilli.

(b) **State whether the following statements are True or False.**

1. *Colletotrichum capsici* is an obligate parasite.
2. *Fusarium solani* belongs to class oomycetes.
3. *Colletotrichum capsici* produce, a minute black dot like fruiting bodies called 'acervuli' .
4. White floury patches symptom appear on both side of the leaves of grape in powdery mildew disease.
5. The acervuli of Gloeosporium do not produce setae.
6. The hyphae of *Colletotrichum. gloeosporioides* accumulate below the host cuticle and develop fruiting bodies called acervuli.
7. Leaf curl of chilli is transmitted by the white fly.

**Fig. 1:** Anthracnose

**Fig. 2:** Wilt

**Fig. 3:** Leaf Curl

**Plate 18.** Photograph showing symptoms of major disease of chilli *(See colour version on page 212)*

# 19

# Diseases of Turmeric and Their Management

| Sl No | Name of Diseases | Causal Organism |
|---|---|---|
| 1 | Leaf spot | *Colletotrichum capsici* Syd. |
| 2 | Leaf blotch | *Taphrina maculans* E. J. Butler |
| 3 | Rhizome rot | *Pythium graminicolum or P. aphanidermatum* (Edson) |

## Leaf spot

### *Colletotrichum capsici*

### Diagnostic Symptoms

- Initially oblong brown spots having about 4-5 cm in length and 2-3 cm in width with grey centres surrounded with yellow hallo are found on the leaves .
- Large number of spots may be found on a single leaf and as the disease advances, spots enlarge and cover a major portion of leaf blade.
- Severly affected plants dry and wilt.
- Black dots acervuli formed in concentric rings on spot.The grey centres become thin and get teared(Fig-2).

### Etiology

### Scientific Classification

Kingdom : Fungi

Division : Eumycota

Sub- division : Deuteromycotina

Class : Coelomycetes

Order : Melanconiales

Family : Melanconiaceae

Genus : *Colletotrichum*

Species : *capsici*

- Mycelium: septate and inter and intra cellular.
- Conidiophores: single celled ,club shaped arise from the hymenial layer below the epidermis, and emerge directly through the epidermis or through stomata.
- Conidia: The conidiophores bear single conidia which are cylindrical to falcate, single celled, hyaline and mostly have blunt ends. They are densely granular, contain oil globules and measure 18-25x3.5-5µ.
- Acervuli: rounded and elongated.
- Setae: brown 1-5 septate, rigid, hardly swollen at the base, slightly tapered towards the paler acute apex.

**Disease cycle**

- Pathogen survives in soil and plant debris.
- Primary infection occurs by spores present in soil
- Secondary infection by conidia through rain or wind.

**Favourable conditions**

- Relative humidity-80%
- Temperature – 21- $25^0$ C
- Leaf wetness.

**Integrated Management**

- Collection and burning of fallen leaves.
- Proper spacing should be maintained.
- Select seed material from disease free areas.
- Crop rotations should be followed.
- Cultivate resistant/ tolerant varieties: Ashwini, Ambika, Angeles, American pride, Surabhi.
- Treat seed material with Mancozeb+Carbendazim @ 2 g/l 1 gm/litre, for 30 minutes and shade dry before sowing.
- Foliar sprays with Carbendazim + Mancozeb @ 0.2%,Blue copper or Blitox 50 @ 0.3 percent , if required repeat 15 days

## Leaf blotch

### *Taphrina maculans*

### Diagnostic Symptoms

- Intial symptom spear as small, oval, rectangular or irregular brown spots on either side of the leaves which soon become dirty yellow or dark brown and leaves also turn yellow.
- In severe cases the plant present a scorched appreance.
- Rhizome yield is reduced(Fig-1).

### Etiology

### Scientific Classification

| | | |
|---|---|---|
| Kingdom | : | Fungi |
| Division | : | Ascomycota |
| Class | : | Taphrinomycetes |
| Order | : | Taphrinales |
| Family | : | Taphrinaceae |
| Genus | : | *Taphrina* |
| Species | : | *maculans* |

- Pathogen hyphae are present in the cuticle and epidermal layers of the host.
- Hausotria: branched or lobed grow into the host cell, drawing out the required nutrients.
- Asci; Cylindrical to clavate, contains eight ascospores .
- Ascospores: Ovoid, hyaline, unicellular, and measure 6-7x2-3μ.They often multiply by budding either in the asci or when released.

### Disease Cycle

- Pathogen survives in soil and plant debris.
- Primary infection occur by spores present in the soil.
- Secondary infection by conidia thorugh rain or wind.

### Favourable Conditions

- High soil moisture.
- Temperature $25^0$ C.
- Leaf wetness.

## Integrated Disease Management

- Proper spacing should be maintained.
- Select seed material from disease free areas.
- Crop rotations should be followed.
- Cultivate resistant/ tolerant varieties: Ashwini, Ambika, Angeles, American pride, Surabhi.
- Treat seed material with Carbendazim + Mancozeb at 2.5 gm/litre of water for 30 minutes and shade dry before sowing.
- Foliar sprays with Blue copper or Blitox 50 @ 0.3percent, if required repeat after 15 days.
- Collection and burning of fallen leaves.

## MODEL QUESTION PAPER

### Section 'A'

### Long answer questions

18. Explain in detail the occurrence, importance, symptoms, etiology, disease cycle and management of rhizome rot of turmeric .
19. Briefly describe the diagnostic symptoms and integrated management of any two of the following turmeric diseases:

    (i) Leaf blotch (ii) Leaf spot

    (iii) Rhizome rot
20. Describe the leaf spot of turmeric in the following headings:

    (i) Pathogen (ii) Symptoms

    (iii) Disease cycle (iv) Disease management
21. Describe in detail the most distinguishing symptoms, causal organism, disease cycle and integrated management of leaf blotch of turmeric .
22. Give the diagrammatic representation of the disease cycle of the following diseases of grapes

    (i) Leaf blotch (ii) Leaf spot

    (iii) Rhizome rot

### Section 'B'

### Short answer questions

1. How will you differentiate leaf spot from leaf blotch of turmeric.
2. Which fungal pathogen causes leaf spot of turmeric? Give its systematic position.

3. Write down an integrated management practices of rhizome rot of turmeirc.
4. Mention etiology of *Colletotrichum capsici* causing leaf blotch of turmeric.
5. Mention favorable condition for rhizome rot of turmeric .

## Section 'D'

### Objective type questions

Fill in the blanks with suitable word (s)

1. Leaf spot of turmeric is caused by ……………………….. ..
2. Leaf blotch of turmeric is caused by ……………
3. The fruiting body formed in leaf spot of turmeric is called ……
4. Foliar spray with metalaxyl fungicide control --------------- disease of turmeric.
5. The asci of *Taphrina maculans* contains ------------ ascospores

### State whether the following statements are True or False

1. *Colletotrichum capsici* is an obligate parasite.
2. Asci of *Taphrina maculans* contains eight ascospores.
3. Setae is present in *Colletotrichum capsici.*
4. An oblong brown spots with grey centres surrounded with yellow hallo is diagnostic symptom of leaf spot of turmeric .
5. Secondary infection takes place by conidia through rain or wind in leaf spot of turmeric.

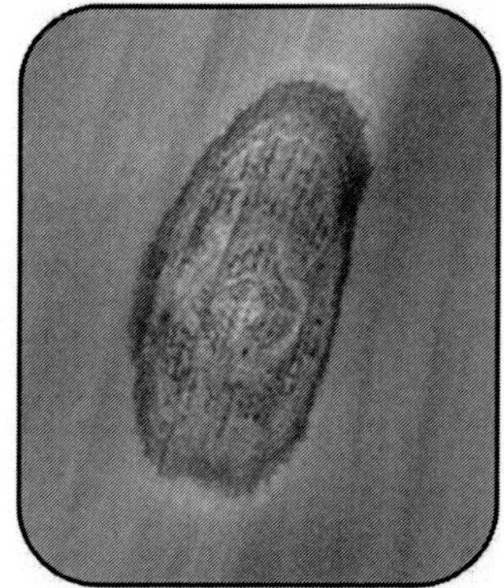

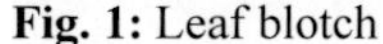

**Fig. 1:** Leaf blotch **Fig. 2:** Leaf spot

**Plate 19.** Photograph showing symptoms of major disease of turmeric *(See colour version on page 212)*

# 20

# Diseases of Coriander and Their Management

| Sl No | Name of Diseases | Causal Organism |
|---|---|---|
| 1 | Stem gall | *Protomyces macrospores* Unger |
| 2 | Powdery mildew | *Eysiphe polygoni* DC |
| 3 | Wilt | *Fusarium oxysporum f.*sp. *corianderii* |

## Stem gall

### Diagnostic Symptoms

- Symptom appears in the form of tumor-like swellings of leaf veins, leaf stalks, peduncles, stems as well as fruits.
- Infected veins show a swollen hanging appearance to the leaves.
- Initially the tumors are glossy which rupture later on and become rough. They are about 3 mm broad and up to 12.5 mm long.
- Badly affected plants may be killed.
- In the presence of excessive soil moisture, especially under shaded conditions, when the stem fails to harden and remain succulent, the tumors are numerous (Fig-1).

### Etiology

### Scientific Classification

Kingdom : Fungi

Division : Ascomycota

Class : Taphrinomycetes

Order : Taphrinales

Family : Protomycetaceae

Genus : *Protomyces*

Species : *macrospores*

- Mycelium: endophytic, intercellular, septate, and irregularly branched. After maturity, some of the hyphal cells swell, become ellipsoidal or globose, and thick walled.
- Cells so transformed represent the resting structures referred to as chlamydospores.
- Chlamydospores: Intercellular and multinucleate mycelium's are formed which are globose to ellipsoid, thick-walled, three-layered membrane and smooth with brownish colour. Mature chlamydospores were multinucleate and measured 60-70 µm x 50-60 µm in diameter.
- Multinucleate protoplast of the chlamydospore migrates to the vesicle. A large vacuole develops in the centre of the vesicle displacing the multinucleate protoplasm to the peripheral region where it undergoes cleavage resulting in its many pieces.
- Protoplasm-pieces are "naked" and each one encloses within it a daughter nucleus.
- Naked cells so produced undergo reduction division that finally results in the formation of four spores inside each naked cell .
- On maturity, these ascospores collect in the centre of the vesicle which, however, bursts and releases them.
- Upon liberation, the ascospores may reproduce by budding in yeast-like manner or they may conjugate in pairs with one member of each pair continuing to bud.
- Conjugation cell (zygote?) apparently germinates and forms a diploid mycelium that penetrates and enters into the host.

## Disease Cycle

- Disease is soil borne.
- Pathogen may survive in soil as resting spore (chlaymydospores) for several years.
- Inoculum present in the soil are the source of primary infection.
- Secondary infection is carried by means of spores.

## Favourable Conditions

- Relative humidity- >85%
- Optimum Temperature – 15- $20^0$ C
- Ph- 7.5

## Integrated Managment

- Deep summer ploughing of fields during summer.
- Use resistant or tolerant varieties.
- Soil solarization: Cover the beds with polythene sheet of 0.45 mm thickness for three weeks before sowing.
- Seed treatment with Carboxin + Thiram @ 2 gm /kg seed.
- Incubate Trichoderma @ 500 g in 100 Kg FYM /acre of the field. before sowing.
- Field sanitation.
- Distroy alternate host plants.
- Balanced applications of manures and fertilizers.
- Follow crop rotation with non host crops like cereals for 3 year.

## MODEL QUESTION PAPER

### Section 'A'

### Long answer questions

1. Explain in detail the occurrence, importance, symptoms, etiology, disease cycle and management of stem gall of coriander.
2. Describe the stem gall of coriander in the following headings:
   (i) Pathogen (ii) Symptoms
   (iii) Disease cycle (iv) Disease management
3. Describe in detail the most distinguishing symptoms, causal organism, disease cycle and integrated management of powdery mildew of coriander.
4. Give the diagrammatic representation of the disease cycle of the following diseases of coriander
   (i) Powdery mildew (ii) Stem gall
   (iii) Wilt

### Section 'B'

### Short answer questions

1. How will you differentiate stem gall from root nodule.
2. Write down the etiology of powdery mildew of coriander.
3. Which fungal pathogen causes stem gall of coriander? Give its systematic

position.

4. Write down the integrated management practices of stem gall of coriander.
5. Describe the causal organisms and mode of penetration of stem gall of coriander.
6. Mention predisposing factors of stem gall of coriander.
7. Which type of disease can be effectively managed by crop rotation.

**Section 'D'**

## Objective type questions

(a) Fill in the blanks with suitable word (s)

1. *Protomyces macrosporus* belongs to family ---------------------------
2. *Protomyces macrosporus* has a complex life cycle including ascospores and ---------------------------
3. The pathogen *Protomyces macrosporus* survive in soil as resting spore called --------------------------------- .
4. Foliar spray with sulfur fungicide control --------------- disease of coriander.
5. The fungus *Protomyces macrosporus* belongs to order ---------------
6. The symptom appears in stem gall of coriander in the form of ------------------like swellings of leaf veins, leaf stalks, peduncles, stems as well as fruits

(b) State whether the following statements are True or False.

1. *Protomyces macrosporus* is an ascomycete fungus.
2. Stem gall of coriander is a soil borne disease.
3. *Protomyces macrosporus* perenates mainly by means of chlaymydospores.
4. White floury patches symptom appear on both side of the leaves of coriander in powdery mildew disease.
5. The life cycle of *Protomyces macrosporus* include ascospores and chlamydospores.

**Fig. 1:** Stem gall

**Fig. 2:** Powdery mildew

**Plate 20:** Photograph showing symptoms of major disease of coriander *(See colour version on page 213)*

# 21

# Diseases of Marigold and Their Management

| Sl No | Name of Diseases | Causal Organism |
|---|---|---|
| 1 | Botrytis gray mould | *Botrytis cinerea* |
| 2 | Fusarium Wilt | *Fusarium oxysporum* |
| 3 | Bacterial leaf spot | *Pseudomonas tagetis* |
| 4 | Leaf Spot | Alternaria spp |

## Botrytis Gray Mould

### Diagnostic Symptoms

- Appearance of a gray, fuzzy mold on flowers, leaves or stems of or plants. Also called, "gray mold".
- Spotting or blight on leaf and flower stem lesions and dieback.
- Flower parts become necrotic and die (Fig-1).

### Etiology

Domain : Eukaryota

Kingdom : Fungi

Division : Ascomycota

Class : Leotiomycetes

Order : Helotiales

Family : Sclerotiniaceae

Genus : *Botrytis*

Species : *cinerea*

Telemorph : *Botryotinia fuckeliana*

- **Colony:** fast-growing, white, low, flaky, becoming grey to brownish grey.

- Conidia: ellipsoidal or ovoid, 6.1 to 8.5 × 5.1 to 9.8 μm and from PDA cultures were 5.1 to 10.5 × 4.3 to 7.2 μm .
- Conidiophores: straight or flexuous, septate, with an inflated basal cell, brown to light brown, and measured 105 to 425 × 9 to 28 μm.
- Sclerotia: black ranging from 0.7 to 4.8 × 1.0 to 3.7 mm.

**Disease Cycle**

- Pathogen overwinter as sclerotia on dead plant debris and as sclerotia or conidia in infested soil.
- Primary infection: Pathogen mycelial strands from previously infected plant parts can grow onto healthy plant parts and infect them.
- Secondary infection: Spores spread by wind or splashing water to infect dying, wounded, or extremely soft plant tissues in the spring.

**Favourable Conditions**

- Moist, humid environment is ideal for pathogen sporulation and spread.
- Optimum temperature for germination of conidia/spore & establishment of infection- 18 to 23$^{0}$C.
- Relative humidity for spore germination-90 to 100 percent.

**Integrated Management**

- Avoid splashing water on the foliage when watering.
- Remove and destroy all infected plant parts as soon as they are observed.
- Give adequate space between plants to allow for good air circulation.
- Avoid fertilizing with excessive amounts of nitrogen.
- Avoid unnecessarily wounding plants.
- Only blemish-free, nonsenescent flowers or plant material should be stored.
- Storage area should be clean, cool, and dry without free moisture on the walls, ceiling, or floor and with a humidity of 90 to 95 percent to prevent shrinking or shriveling of plant material. The temperature should be as close to freezing.
- Foliar spray with Carbendazim + Mancozeb @ 0.2% or Hexaconzole + Zineb @ 0.2%,repeat after 15 days.
- Biological controls have been successfully used in the form of fungi-like *Trichoderma harzianium* Rifai to control gray mold.

## MODEL QUESTION PAPER

### Section 'A'

### Long answer questions

1. Explain in detail the occurrence, importance, symptoms, etiology, disease cycle and management of Alternaria Leaf Spot of marigold.
2. Describe in detail the most distinguishing symptoms, causal organism, disease cycle and integrated management of botrytis gray mould of marigold

### Section 'B'

### Short answer questions

1. Write down the etiology of botrytis gray mould of marigold.
2. Which fungal pathogen causes botrytis gray mould of marigold? Give its systematic position.
3. Write down the integrated management practices of botrytis gray mould of marigold.

### Section 'D'

### Objective type questions

(a) Fill in the blanks with suitable word (s)

1. Botrytis gray mould of marigold is caused by ..........................
2. The telemorphic stage of botrytis gray mould of marigold is ......
3. *Botrytis cinerea* is a ------------------------- fungus .
4. The pathogen *Botrytis cinerea* causes botrytis gray mould of marigold overwinter as----------------------------------in infested soil.
5. ------------------------- is a fungal parasite of *Botrytis cinerea*

(b) State whether the following statements are True or False.

1. *Botrytis cinerea* ia an obligate parasite.
2. *Botrytis cinerea* belongs to class oomycetes.
3. Appearance of a gray, fuzzy mold on flowers, leaves or stems of marigold is called gray mold.
4. The colony of *Botrytis cinerea* is fast-growing and white.
5. *Gliocladium roseum* is a fungal parasite of *Botrytis cinerea.*

6. Different *Botrytis cinerea* strains show considerable genetic variability.
7. The conidia of *Botrytis cinerea* dispersed by wind and by rain-water, cause new infections.
8. *Botrytis cinerea* is a necrotrophic fungus that affects many plant species.
9. Excessive application of nitrogen will increase the incidence of disease while not improving yields.
10. *Botrytis cinerea* produces highly resistant sclerotia as survival structures in older cultures.

**Fig. 1:** Botrytis grey mould

**Fig. 2:** Powdery mildew

**Plate 21.** Photograph showing symptoms of major disease of marigold *(See colour version on page 213)*

# 22

# Diseases of Rose and Their Management

| Sl No | Name of Diseases | Causal Organism |
|---|---|---|
| 1 | Powdery mildew | *Sphaerotheca pannosa var. rosae* |
| 2 | Die-back | *Diplodia rosarum* |
| 3 | Black spot | *Diplocarp an rosae* |
| 4 | Rust | *Phragmidium mucronatum* |

## Powdery mildew

### *Sphaerotheca pannosa* var. *rosae (Sphaerotheca pannosa)*

### Diagnostic Symptoms

- A white, powdery fungal growth on the upper, lower or both leaves and shoots.
- There may be discolouration of the affected parts of the leaf, and heavily infected young leaves can be curled and distorted.
- Mildew growth may also be found on the stems, flower stalks, calyces and petals.
- Heavily infected flower buds frequently fail to open properly.
- Mildew growth on stems and flower stalks is usually thicker and more mat-like than that on the leaves
- Mildew growth on all parts may turn browner as it ages (Fig-1).

### Etiology

### Scientific Classification

Kingdom : Fungi

Division : Eumycota

Class : Leotiomycetes

Order : Erysiphales

Family : Erysiphaceae

Genus : *Sphaerotheca*

*Species* : *pannosa*

- Obligate parasite: Pathogen must have a living host complete its life cycle, which can be as short as 72to 96 hours in favouable conditions.
- Mycelium: white, septate, ectophytic and sends globose haustoria into the epidermal cells of the host.
- Conidiophores: are short and erect.
- Conidia: one celled, oblong, minutely verrucose with many large fat globules.
- Cleistothecia: formed towards the end of the season on the leaves, petals, stems and thorns with simple myceloid appendages.
- Ascus: contains eight ascospores

## Disease Cycle

- Pathogen survives as mycelium in dormant buds and shoots or as cleistothecia.
- They discharge their spores in the spring.
- Secondary infection occurs through wind borne conidia throughout year.
- However, when it grows it can survive in buds, emerging with the appearance of the flowers.

## Favourable Conditions

- Hot and warm days with cool nights.
- Water on leaf surface prevents the spores from germinating.

## Integrated Management

- Collection and burning of fallen leaves.
- Prune the infected plants and dispose of them properly.
- Provide plants with adequate nutrients and water to maintain their immune defenses.
- Keep the soil well watered and mulched to prevent moisture loss and to cover up overwiterinng spores.
- Maintain proper spacing to provide good air circulation and prune them regularly to prevent overcrowding.
- Use fan to provide adequate ventilation during humid nights.

- Resistant varieties: Ashwini, Ambika, Angeles, American pride, Surabhi
- Foliar spray with combination fungicide *viz.*, Captan + Hexaconazole WP @ 0.3% or Azoxystrobin + Tebuconazole @ 0.1% or Tebuconazole with Trifloxystrobin @ 0.1%, if required repeat after 15 days.

## Die-back

***Diplodia rosarum/Botryodiplodia theobromae***

### Diagnostic Symptoms

- Browning and dieback of the tips of young shoots in spring.
- Browning and dieback of a pruning stub, which then progresses further down the branch
- Dieback of twigs, branches, main stems or even the whole plant at any time of year
- Fungal structures, such as tiny black fruiting bodies, are sometimes visible on the affected parts of the plant
- In some instances there may also be associated root decay (Fig-2).

### Etiology

### Scientific Classification

Kingdom : Fungi

Division : Ascomycetes

Class : Dothideomycetes

Order : Botriosphaeriales

Family : Botriosphaeiaceae

Genus : *Diplodia*

*Species : rosarum*

- Pathogen produces round, black pycnidia which bear spores.
- Pycnidia: round, black which bear spores.
- Pycnidiospores: dark coloured and two celled.
- Perithecia: immersed in the host tissue and are surrounded by a pseudostroma.
- Ascospores: ellipsoidal or fusoid, hyaline, two celled with the septum in or near the middle.

## Disease Cycle

- Pathogen survives in plant debris.
- Spores from pycnidia are considered as primary source of infection and spread.
- Scondary infection takes place thorugh spores spread through air current or blowing rain.

## Favourable Conditions

- High humidiy
- Temperature-30-32$^0$C.

## Integrated Management

- Plant roses in well-prepared soil, making sure that the roots are well spread out
- Do not plant roses into soil that has grown roses previously, without taking remedial action. Change the soil to a depth of at least 30cm and a width of at least 60cm.
- Use resistant / tolerant varieties: Blue moon, Red gold, Summer queen, etc.
- Pruning should be done so that lesions on the young shoots will be eliminated and apply chaubatia paste in the pruned area.
- Avoid soils that are prone to either drought or waterlogging
- Feed plants in spring with a proprietary rose fertiliser and mulch the soil to prevent water loss
- Foliar spray with Captan + Hexaconazole WP @ 0.3% or Azoxystrobin + Tebuconazole @ 0.1% or Tebuconazole with Trifloxystrobin @ 0.1%, if required repeat after 15 days.

## Black spot

***Diplocarpan rosae (Anamorph: Marssonina rosae)***

## Diagnostic Symptoms

- Dark brown tar coloured spots with fringed borders.
- Spots coalesce forming large patches.
- Infected leaves turn brown and defoliate.
- Fungus may also attack stems and flowers of rose bushes (Fig-3).
- On stems, infected areas are blackened with blistered appearance dotted with pustules.

## Etiology

### Scientific Classification

Kingdom : Fungi

Division : Ascomycota

Class : Leotiomycetes

Order : Helotiales

Family : Dermateaceae

Genus : *Diplocarpan*

Species : *rosae*

- Vegetative body of the fungus consists of two parts *viz*., the subcuticular mycelium and the internal mycelium.
- Fungus produces acervuli between the outer wall and cuticle of the epidermis.
- Fungus produces *Marssonina*-type conidia in acervuli and ascospores in tiny apothecia formed in old lesions.
- Conidiophore: short arise from a thin black stroma and give rise to successive crops of conidia.
- Conidia push up and rupture the cuticle.

### Disease Cycle

- Pathogen survives as mycelia, ascospores and conidia in infected leaves and canes.
- Both kinds of spores can cause primary infections of leaves in the spring by direct penetration. The mycelium grows in the mesophyll, but within two weeks forms acervuli and conidia at the upper surface.
- Conidia are produced throughout the growing season and cause repeated infections during warm, wet weather.

### Integrated Management

- Use resistant / tolerant varieties: Bebebune, Coronado, Grand opera, Sphinx.
- Affected parts should be collected and destroyed.
- Space roses far enough apart for good air circulation.
- Spray Tridemorph @0.25% or Cholorothalonil @ 2ml/L of water at weekly intervals starting with the sprouting of the plants till new foliage appears.

- Avoid overhead watering and keep foliage as dry as possible.
- Radiance - escape infection due to waxy surface.

## MODEL QUESTION PAPER

### Section 'A'

### Long answer questions

1. Explain in detail the occurrence, importance, symptoms, etiology, disease cycle and management of powdery mildew of rose .
2. Briefly describe the diagnostic symptoms and integrated management of black spot and dieback diseases of rose.
3. Describe the black spot of rose in the following headings:

   (i) Pathogen (ii) Symptoms

   (ii) Disease cycle (iv) Disease management
4. Describe in detail the most distinguishing symptoms, causal organism, disease cycle and integrated management of die back of rose.
5. Give the diagrammatic representation of the disease cycle of the following diseases of rose

   (iii) Powdery mildew (ii) Die back

   (iii) Black spot

### Section 'B'

### Short answer questions

1. Write down the etiology of powdery mildew of rose.
2. Which fungal pathogen causes powdery mildew of rose? Give its systematic position.
3. Write down the integrated management practices of black spot diseases of rose.
4. Suggest some chemicals for management of powdery mildew of rose.
5. Describe favourable condition for development of black spot and dieback disease of rose. .

## Section 'D'

## Objective type questions

(a) Fill in the blanks with suitable word (s)

1. Powdery mildew of rose is caused by ……………………….. ..
2. Black spot of rose is caused by ……………
3. The mycelium *of Diplocarpan rosae* forms ----------- and conidia at the upper surface within two weeks of infection
4. Foliar spray with sulfur fungicide control --------------- disease of rose
5. *The Diplocarpan rosae* produces *Marssonina*-type conidia in acervuli and ascospores in tiny fruiting body called -----------------------.
6. The telemorphic stage of *Diplocarpan rosae is* ---------------------
7. The perithecia of *Diplodia rosarum* immersed in the host tissue and are surrounded by a ----------------.

(b) State whether the following statements are True or False.

1. *Diplodia rosarum* produces round, black pycnidia which bear spores.
2. *Diplocarpan rosae* survives as mycelia, ascospores and conidia in infected leaves and canes.
3. The pathogen *Sphaerotheca pannosa* is an obligate parasite.
4. White floury patches symptom appear on both side of the leaves of rose in powdery mildew disease.
5. The fungus *Diplocarpan rosae* produces acervuli between the outer wall and cuticle of the epidermis.

**Fig. 1:** Powdery mildew

**Fig. 2:** Die back

**Fig. 3:** Black spot

**Plate 22:** Photograph showing symptoms of major diseases of rose *(See colour version on page 214)*

# Colour Plates

**Fig. 1:** Loose smut

**Fig. 2:** Stem rust

**Fig. 3:** Brown rust

**Fig. 4:** Yellow rust

**Fig. 5:** Karnal bunt

**Fig. 6:** Leaf blight

**Plate 1:** Photograph showing symptoms of major diseases of wheat
(*See black & white version on page 28*)

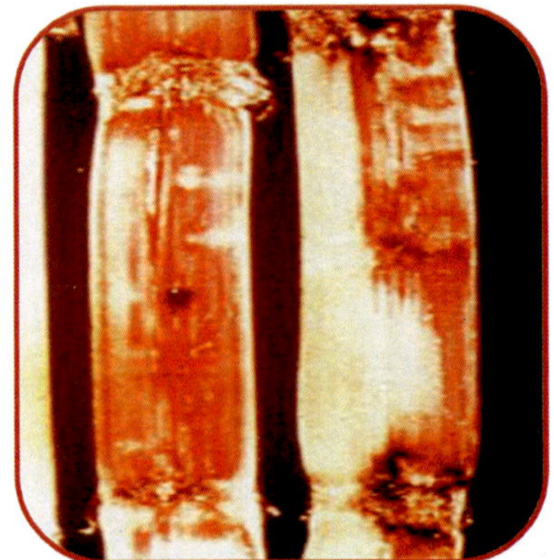

**Fig. 1:** Red rot

**Fig. 2:** Smut

**Fig. 3:** Grassy shoot

**Fig. 4:** Ratton stuting

**Fig. 5:** Pokkahboeng

**Fig. 6:** Wilt

**Plate-2.** Photograph showing symptoms of major diseases of Sugarcane (*See black & white version on page 40*)

**Fig. 1:**.Alternaria leaf blight

**Fig. 2:** Sclerotinia rot

**Plate 3:** Photograph showing symptoms of major diseases of Sunflower
(*See black & white version on page 46*)

**Fig. 1:** White rust

**Fig. 2:** Downey mildew

**Fig. 2:** Alternaria leaf spot

**Fig. 4:** Sclerotinia Stem rot

**Plate 4:** Photograph showing symptoms of major diseases of mustard
(*See black & white version on page 57*)

**Fig. 1:** Ascochyta blight

**Fig. 2:** Wilt

**Fig. 3:** Botrytis grey mold

**Plate 5:** Photograph showing symptoms of major diseases of chickpea
(*See black & white version on page 66*)

**Fig. 1:** Wilt

**Fig. 2:** Rust

**Plate 6:** Photograph showing symptoms of major diseases of lentil
(*See black & white version on page 72*)

**Fig. 1:** Fusarium wilt

**Fig. 2:** Anthracnose

**Fig. 3:** Bacterial blight

**Plate 7:** Photograph showing symptoms of major diseases of cotton (*See black & white version on page 81*)

**Fig. 1:** Powdery mildew

**Fig. 2:** Downy mildew

**Fig. 3:** Rust

**Plate 8:** Photograph showing symptoms of major diseases of pea
(*See black & white version on page 90*)

**Fig. 1:** Powdery mildew

**Fig. 2:** Anthracnose

**Fig. 3:** Bacterial black spot

**Fig. 4:** Flower malformation

**Plate 9:** Photograph showing symptoms of major diseases of mango
(*See black & white version on page 102*)

**Fig. 1:** Gummosis

**Fig. 2:** Canker

**Plate 10:** Photograph showing symptoms of major diseases of citrus
(*See black & white version on page 108*)

**Fig. 1:** Powdery mildew

**Fig. 2:** Downy mildew

**Fig. 3:** Anthracnose

**Plate 11:** Photograph showing symptoms of major diseases of grape
(*See black & white version on page 117*)

**Fig. 1:** Scab

**Fig. 2:** Powdery mildew

**Fig. 3:** Crown gall

**Fig. 4:** Fire blight

**Plate 12:** Photograph showing symptoms of major diseases of apple (*See black & white version on page 130*)

**Fig. 1:** Leaf curl

**Plate 13:** Photograph showing symptoms of major disease of peach (*See black & white version on page 134*)

**Fig. 1:** Leaf Spot

**Plate 14:** Photograph showing symptoms of major disease of Strawberry (*See black & white version on page 138*)

**Fig. 1:** Late blight

**Fig. 2:** Early blight

**Fig. 3:** Black scurf

**Fig. 4:** Mosaic

**Fig. 5:** Leaf curl

**Plate 15:** Photograph showing symptoms of major disease of potato
(*See black & white version on page 151*)

**Fig. 1:** Powdery mildew

**Fig. 2:** Downy mildew

**Fig. 3:** Fusarium wilt

**Fig. 4:** Anthracnose

**Plate 16:** Photograph showing symptoms of major diseases of cucurbits
(*See black & white version on page 162*)

**Fig. 1:** Purple blotch

**Fig. 2:** Stemphyllium blight

**Plate 17:** Photograph showing symptoms of major disease of onion
(*See black & white version on page 167*)

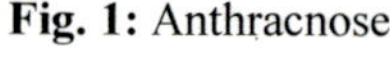

**Fig. 1:** Anthracnose

**Fig. 2:** Wilt

**Fig. 3:** Leaf Curl

**Plate 18.** Photograph showing symptoms of major disease of chilli
(*See black & white version on page 175*)

**Fig. 1:** Leaf blotch

**Fig. 2:** Leaf spot

**Plate 19.** Photograph showing symptoms of major disease of turmeric
(*See black & white version on page 181*)

**Fig. 1:** Stem gall

**Fig. 2:** Powdery mildew

**Plate 20:** Photograph showing symptoms of major disease of coriander
(*See black & white version on page 187*)

**Fig. 1:** Botrytis grey mould

**Fig. 2:** Powdery mildew

**Plate 21.** Photograph showing symptoms of major disease of marigold
(*See black & white version on page 192*)

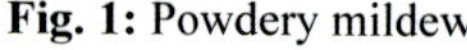

**Fig. 1:** Powdery mildew

**Fig. 2:** Die back

**Fig. 3:** Black spot

**Plate 22:** Photograph showing symptoms of major diseases of rose
(*See black & white version on page*)